Oxford Lecture Series in
Mathematics and its Applications 26

Series editors
John Ball   Dominic Welsh

OXFORD LECTURE SERIES
IN MATHEMATICS AND ITS APPLICATIONS

Dynamics of Viscous Compressible Fluids

Eduard Feireisl

Mathematical Institute of the Academy of Sciences of the Czech Republic

OXFORD
UNIVERSITY PRESS

OXFORD
UNIVERSITY PRESS

Great Clarendon Street, Oxford OX2 6DP

Oxford University Press is a department of the University of Oxford.
It furthers the University's objective of excellence in research, scholarship,
and education by publishing worldwide in

Oxford New York

Auckland Bangkok Buenos Aires Cape Town Chennai
Dar es Salaam Delhi Hong Kong Istanbul Karachi Kolkata
Kuala Lumpur Madrid Melbourne Mexico City Mumbai Nairobi
São Paulo Shanghai Taipei Tokyo Toronto

Oxford is a registered trade mark of Oxford University Press
in the UK and in certain other countries

Published in the United States
by Oxford University Press Inc., New York

A catalogue record for this title is available from the British Library

Library of Congress Cataloging in Publication Data
(Data available)
ISBN 0 19 852838 8 (Hbk)

10 9 8 7 6 5 4 3 2 1

Typeset by Newgen Imaging Systems (P) Ltd., Chennai, India
Printed in Great Britain
on acid-free paper by
Biddles Ltd. www.biddles.co.uk

CONTENTS

PREFACE

Since the time of Euler, partial differential equations have been used to describe the movement of continuous media like fluids or vibration of solids. In its origins, hydrodynamics was a largely mathematical science for which much of the theories of partial differential equations of the eighteenth century were developed. As applications and experimental studies grew more numerous enhanced even more by a spectacular development of modern supercomputers, the rigorous mathematical theory gave ground to numerical studies and computer experiments commonly called applied mathematics of recent days. The Navier–Stokes equations are the most frequently used as a model of linearly viscous incompressible fluids (liquids) while the Euler equations play the same role for inviscid incompressible or compressible fluids (gases). Given the number of numerical results and successful practical applications, it comes as a striking fact how much less is known about the solutions to these equations at a purely theoretical level. The existence (or non-existence) of global-in-time regular solutions to the incompressible Navier–Stokes equations when the ambient physical space is three dimensional is one of the most challenging open problems of the modern theory of partial differential equations. The Euler equations of a compressible fluid form a nonlinear hyperbolic system for which any large data existence problem seems widely open even in the class of distributional solutions. Moreover, fluids modeled by these equations may exhibit very complicated chaotic or self-organized structures commonly denoted as "turbulent" phenomena, the understanding of which represents one of the main challenges of modern mathematical physics. It is hardly conceivable that any real progress in this direction could be made without answering the basic questions of well-posedness of the underlying equations— existence, uniqueness, stability, and continuous dependence of solutions on the data.

This book is designed as a contribution to the mathematical theory of viscous compressible fluids. In accordance with the basic principles of classical continuum mechanics, the state of a fluid at a given time is fully characterized by three macroscopic quantities—the density, the velocity, and the temperature. These satisfy a system of partial differential equations expressing the conservation of mass, momentum, and energy. In the situation when the only source of internal energy change is of purely mechanical origin; that is, when both conduction of heat and its generation by dissipation of mechanical energy can be neglected, the temperature changes can be expressed in terms of the density, and the original system reduces to the Navier–Stokes equations of a compressible barotropic fluid governing the time evolution of the density and momentum. In this model, the pressure of the fluid is given as an explicit function of the density. One can go even further assuming the fluid is incompressible, which amounts to the hypothesis that the density is constant. The resulting system of the incompressible Navier–Stokes equations contains the velocity as the only unknown variable and represents probably the best known model problem in mathematical fluid dynamics.

The existence of global-in-time weak solutions for the incompressible Navier–Stokes equations was established by Leray. His notion of weak solution (1934) preceded both the introduction of the Sobolev spaces (1936) and the generalized derivatives (Schwartz 1944). A comparable theory for viscous compressible baro-tropic fluids has been developed only recently by P.-L. Lions in "Mathematical topics in fluid dynamics, II", Oxford University Press 1998. The major discovery made by Lions are some unexpected properties of a quantity commonly termed as effective viscous pressure. In spite of being strongly nonlinear, the effective viscous pressure behaves almost like it was a weakly continuous function of the density. Such a behavior is of course reminiscent of the quantities studied in the theory of compensated compactness but looking for a straightforward explanation in terms of this theory would be more misleading than elucidating. Another important ingredient of Lions' approach is the concept of renormalized solution developed in the framework of a joint programme with DiPerna (1989).

The major thrust of this book is to develop further the mathematical theory of viscous compressible fluids pursuing two main goals:

- Global existence results for the full system of the Navier–Stokes equations with large data supplemented with a suitable set of constitutive equations.
- Optimal existence results for the barotropic flows with respect to the available *a priori* estimates.

To this end, we introduce two new tools: (i) an oscillations defect measure—to obtain a more precise description of possible oscillations of the density component in a sequence of (approximate) solutions; (ii) a renormalized limit of a sequence of bounded integrable functions—to cope with possible concentrations in the temperature.

The material is organized in the following manner. Chapter 1 is devoted to a review of the underlying physical theory. Besides the basic notions of reference configurations, kinematics, constitutive equations, and balance laws, this part includes an account of general pressure–density–temperature relations including those arising in low energy nuclear physics and astrophysics. In particular, some examples of non-monotone pressure–density constitutive laws are presented.

The basic mathematical concepts used in the book are resumed in Chapters 2 and 3. In order to underline the physical background of the function spaces, the exact definitions introduced in Chapter 2 are followed by the energy estimates deduced in Chapter 3 directly for any smooth solution of the underlying equations (*a priori* estimates). Several other fundamental concepts are also treated at length: average continuity of weak solutions, renormalized solutions of the continuity equation, and instantaneous values of the state variables.

The concept of a variational solution is introduced in Chapter 4. Each equation of the full system, that means, the equation of continuity, the momentum equation, and the energy equation, is treated separately. The basic facts of the theory of renormalized solutions of the continuity equation are reviewed. Moreover, the renormalized solutions of the thermal energy are introduced as a suitable tool to deal with possible concentrations of temperature. The discussion on the variational solutions is further developed in Chapter 5, where more delicate estimates of the pressure and temperature necessary for future analysis are obtained.

Chapter 6 is central to the book. Given the rather poor *a priori* estimates available, the methods of weak convergence play a decisive role in the mathematical theory to be developed in this book. Both "classical" problems of this approach—the presence of oscillations and concentrations in sequences of approximate solutions—are present. The well-known results of the theory of compensated compactness are used in order to cope with possible density oscillations. More specifically, the fundamental properties of the effective viscous pressure discovered by P.-L. Lions are discussed together with an alternative proof of "weak continuity" of this quantity via the famous div-curl lemma. Next, the concept of oscillation defect measure is introduced, and its relation to the propagation of oscillations and the renormalized continuity equation is established. Furthermore, the whole machinery is applied to the crucial problem of propagation of density oscillations in a sequence of solutions, in particular, it is shown that the oscillations decay in time at a uniform rate independent of the choice of initial data provided the pressure is a monotone function of the density. The weak sequential stability (compactness) of the set of weak solutions is established for optimal values of the "adiabatic" exponent. In particular, the physically interesting case of the monoatomic gas in the isentropic regime in three space dimensions can be treated—a problem left open in the up to now available theory. Possible concentrations in the temperature are treated via the method of renormalization (rescaling). A "renormalized" formulation of the thermal energy equation introduced in Chapter 4 is supplemented with the concept of a renormalized limit, usefulness of which being demonstrated on the problem of weak sequential stability and the study of possible concentrations of the temperature in the thermal energy equation.

Chapter 7 contains a complete proof of the existence of global-in-time variational solutions for the full system of the Navier–Stokes equations of a viscous compressible and heat conducting fluid under suitable restrictions imposed on the constitutive laws. These restrictions are by no means optimal but, on the other hand, they seem to be in a good agreement with the underlying physical theory. Probably the most questionable hypothesis seems to be the necessity of the viscosity coefficients to be constant in all temperature regimes. On the other hand, as a byproduct of our approach, we derive "optimal" existence results for the barotropic flows with respect to the available (known) *a priori* estimates.

Another novelty allowed by the present method is the possibility to consider general, not necessarily monotone, pressure–density constitutive equations arising in applications.

This book is intended to be a compact and self-contained presentation of the most recent results of the mathematical theory of viscous compressible fluids including some applications to more specific problems. In order to place the text in better perspective, each chapter is concluded with a section of historical notes including references to all important and relatively new results. However, the cited works have been chosen on the basis of selectiveness rather than completeness to keep the bibliography concise. The results presented in the book are by no means the last word on the subject but they rather indicate possible directions of future research.

E. Feireisl
September, 2003

ACKNOWLEDGEMENT

The work of the author was supported by Grant 201/02/0854 of GA ČR. He is also indebted to Sonja Goj and Josef Málek for careful reading of the manuscript.

1

PHYSICAL BACKGROUND

An investigation into the behavior of a *fluid* in motion may be undertaken from either a *microscopic* or a *macroscopic* point of view. If we attempt to use the microscopic description, the position of each atom (molecule) of the fluid as well as its velocity at a given time must be specified. In order to capture completely the behavior of such a system, it is necessary to deal with a large number of equations describing the motion of each individual particle.

The macroscopic approach pursued in this book reduces the number of variables to a few related to the average effects of action of many molecules. These effects are perceived by our senses and can be measured by instruments.

From the macroscopic point of view, a fluid is always understood as a *continuum* occupying at a given time $t \in R$ a certain spatial domain Ω in the N-dimensional *Euclidean space* R^N. The *state* of the fluid is identified through certain observable macroscopic properties such as the *density* ϱ, the *velocity field* $\mathbf{u}$, and the *temperature* ϑ. These quantities will be assumed to have the same value for a given state regardless of how the system arrived at that state. Accordingly, the time evolution of a fluid is described through a system of partial differential equations, with the *time* $t \in R$ and the *spatial position* $\mathbf{x} \in \Omega$ as independent variables, and the state functions $\varrho = \varrho(t, \mathbf{x})$, $\mathbf{u} = \mathbf{u}(t, \mathbf{x})$, and $\vartheta = \vartheta(t, \mathbf{x})$ as unknowns.

1.1 Kinematics, description of motion

1.1.1 Motions

A *motion* of a body in continuum mechanics is described by a family of one-to-one mappings

$$\mathbf{X}(t, \cdot) : \Omega \to \Omega, \quad t \in I,$$

where $I \subset R$ is a *time interval*, and $\Omega \subset R^N$ is a spatial domain occupied by the body. *Continuum hypothesis* requires $\mathbf{X}(t, \cdot)$ to be a diffeomorphism for any fixed time $t \in I$. It is convenient to choose a *reference configuration* $\mathbf{X}(t_I, \mathbf{x}) = \mathbf{x}$ for all $\mathbf{x} \in \Omega$ at a certain time $t_I \in I$. Accordingly, the curve $\mathbf{X}(t, \mathbf{x})$, $t \in I$, represents the trajectory of a particle occupying at the time t_I a spatial position $\mathbf{x} \in \Omega$. Thus the motion is visualized as mapping parts of space onto parts of space. The reference configuration is introduced to allow us to use *Euclidean geometry* of the ambient space.

There are several ways to describe a motion but all of them are equivalent provided the motion is smooth. Since we are concerned with the dynamics of fluids, it is convenient to use the *spatial description*, usually called *Eulerian*, where the time $t \in I$ and the place $\mathbf{x} \in \Omega$ in the ambient space play the role of independent variables.

A smooth motion $\mathbf{X}$ is completely determined by a *velocity field* $\mathbf{u} : I \times \Omega \to R^N$ through a system of equations:

$$\frac{\partial \mathbf{X}(t, \mathbf{x})}{\partial t} = \mathbf{u}(t, \mathbf{X}(t, \mathbf{x})), \qquad \mathbf{X}(t_I, \mathbf{x}) = \mathbf{x} \quad \text{for } \mathbf{x} \in \Omega, \ t \in I. \tag{1.1}$$

Applying the spatial *gradient operator* $\nabla_x \equiv (\partial_{x_1}, \ldots, \partial_{x_N})$ to both sides of (1.1) we get

$$\frac{\partial \nabla_x \mathbf{X}(t, \mathbf{x})}{\partial t} = \nabla_x \mathbf{u}(t, \mathbf{X}(t, \mathbf{x})) \nabla_x \mathbf{X}(t, \mathbf{x}), \quad \nabla_x \mathbf{X}(t_I, \cdot) = \mathbb{I},$$

from which we deduce

$$\frac{\partial}{\partial t} [\det \nabla_x \mathbf{X}(t, \mathbf{x})] = \text{div}_x \mathbf{u}(t, \mathbf{X}(t, \mathbf{x})) [\det \nabla_x \mathbf{X}(t, \mathbf{x})],$$

where $\text{div}_x \mathbf{u} \equiv \text{trace}[\nabla_x \mathbf{u}]$ denotes the *divergence operator* (cf. Chapter 1 of [117]).

The same relation can be written in an equivalent form as a *transport equation*

$$\partial_t J + \text{div}_x (J\mathbf{u}) = 2J \text{div}_x \mathbf{u} \quad \text{in } I \times \Omega \tag{1.2}$$

for the quantity $J(t, \mathbf{X}(t, \mathbf{x})) \equiv |\det \nabla_x \mathbf{X}(t, \mathbf{x})|$ termed *specific volume*.

1.1.2 Mass transport

In general, *mass* may be thought of as a family of non-negative measures $\{M_t\}$, $t \in I$, on Ω obeying the *principle of mass conservation*:

$$M_{t_1}[\mathbf{X}(t_1, B)] = M_{t_2}[\mathbf{X}(t_2, B)] \tag{1.3}$$

for any Borel set $B \subset \Omega$ and t_1, $t_2 \in I$. Accordingly, the distribution of mass at each time $t \in I$ is uniquely determined by the motion $\mathbf{X}$ and the reference distribution M_{t_I}.

We shall assume that M is absolutely continuous with respect to the standard Lebesgue measure; that means the mass distribution is characterized by a *density* function $\varrho = \varrho(t, \mathbf{x})$, which is non-negative and locally integrable on Ω.

Consequently, the relation (1.3) may be rewritten as

$$\int_{\mathbf{X}(t_1, B)} \varrho(t_1, \mathbf{x}) d\mathbf{x} = \int_{\mathbf{X}(t_2, B)} \varrho(t_2, \mathbf{x}) \, d\mathbf{x} \quad \text{for any } t_1, t_2 \in I$$

or, equivalently,

$$\frac{d}{dt} \int_{\mathbf{X}(t, B)} \varrho(t, \mathbf{x}) \, dx = 0.$$

If the motion is smooth, the same equation may be expressed with the help of the *convection theorem* (see Theorem 3.1 of Chapter 1 in [95]) in the form:

$$\frac{d}{dt} \int_B \varrho(t, \mathbf{x}) d\mathbf{x} + \int_{\partial B} \varrho(t, \mathbf{x}) \, \mathbf{u}(t, \mathbf{x}) \cdot \mathbf{n} \, d\sigma = 0 \qquad (1.4)$$

for any bounded domain $B \subset \Omega$ with a sufficiently smooth boundary ∂B that the outer normal vector $\mathbf{n} = \mathbf{n}(\mathbf{x})$ may be defined for any $\mathbf{x} \in \partial B$.

If the density ϱ is smooth, one can use Green's theorem to deduce the *continuity equation*:

$$\partial_t \varrho + \operatorname{div}_x(\varrho \mathbf{u}) = 0 \quad \text{in } I \times \Omega, \qquad (1.5)$$

which is a mathematical formulation of the physical *principle of mass conservation*.

If ϱ is strictly positive, division of equation (1.5) by ϱ^2 yields

$$\partial_t \varrho^{-1} + \operatorname{div}_x(\varrho^{-1} \mathbf{u}) = 2\varrho^{-1} \operatorname{div}_x \mathbf{u}$$

which is nothing other than equation (1.2). This yields a relation between the fluid density ϱ and the *specific volume J*, namely,

$$J(t, \mathbf{x}) = \frac{\varrho(t_I, \mathbf{X}^{-1}(t, \mathbf{x}))}{\varrho(t, \mathbf{x})} \quad \text{for } t \in I, \ \mathbf{x} \in \Omega.$$

In other words, equations (1.2) and (1.5) are equivalent provided both ϱ and J are smooth, and ϱ strictly positive. The regions of zero density are usually associated with regions of cavitation, the density being positive elsewhere.

1.2 Balance laws

1.2.1 Equations of motion

The inertial nature of a fluid is expressed when we apply to a small element *Newton's second law of motion* to obtain

$$\frac{d}{dt} \int_{\mathbf{X}(t, B)} (\varrho \mathbf{u})(t, \mathbf{x}) \, d\mathbf{x} = \int_{\mathbf{X}(t, B)} \varrho(t, \mathbf{x}) \mathbf{f}(t, \mathbf{x}) \, d\mathbf{x} + \int_{\partial \mathbf{X}(t, B)} \mathbf{t}(t, \mathbf{x}, \mathbf{n}) \, d\sigma, \qquad (1.6)$$

or, by virtue of the convection theorem,

$$\frac{\mathrm{d}}{\mathrm{d}t} \int_B (\varrho \mathbf{u})(t, \mathbf{x}) \, \mathrm{d}\mathbf{x} + \int_{\partial B} (\varrho \mathbf{u})(t, \mathbf{x}) \, \mathbf{u}(t, \mathbf{x}) \cdot \mathbf{n} \, \mathrm{d}\mathbf{x}$$

$$= \int_B \varrho(t, \mathbf{x}) \, \mathbf{f}(t, \mathbf{x}) \, \mathrm{d}\mathbf{x} + \int_{\partial B} \mathbf{t}(t, \mathbf{x}, \mathbf{n}) \, \mathrm{d}\sigma \qquad (1.7)$$

where the right-hand side is the resultant force acting on a volume element $B \subset \Omega$. In accordance with the *Euler–Cauchy stress principle*, the resultant force can be written as the sum of an *external force* with density $\mathbf{f}$ and a *simple traction* represented by the vector $\mathbf{t}$.

The stress principle in continuum mechanics is put to use through two fundamental *laws of Cauchy* (see e.g. Section 2.6 of Chapter 2 in [117]):

- There is a *stress tensor* $\mathbb{T} = \mathbb{T}(t, \mathbf{x})$ such that

$$\mathbf{t}(t, \mathbf{x}, \mathbf{n}) = \mathbb{T}(t, \mathbf{x})\mathbf{n}.$$

- The stress tensor $\mathbb{T}$ is symmetric.

Accordingly, equation (1.7) takes the form

$$\frac{\mathrm{d}}{\mathrm{d}t} \int_B (\varrho \mathbf{u})(t, \mathbf{x}) \, \mathrm{d}\mathbf{x} + \int_{\partial B} (\varrho \mathbf{u})(t, \mathbf{x}) \mathbf{u}(t, \mathbf{x}) \cdot \mathbf{n} \, \mathrm{d}\mathbf{x}$$

$$= \int_B \varrho(t, \mathbf{x}) \, \mathbf{f}(t, \mathbf{x}) \, \mathrm{d}\mathbf{x} + \int_{\partial B} \mathbb{T}(t, \mathbf{x})\mathbf{n} \, \mathrm{d}\sigma. \qquad (1.8)$$

If all quantities are smooth, we can apply Green's theorem to obtain the *momentum equation*:

$$\boxed{\partial_t(\varrho \mathbf{u}) + \mathrm{div}_x(\varrho \mathbf{u} \otimes \mathbf{u}) = \mathrm{div}_x \mathbb{T} + \varrho \mathbf{f} \quad \text{in } I \times \Omega,} \qquad (1.9)$$

where the symbol $\otimes$ stands for the *tensor product*, $[\mathbf{u} \otimes \mathbf{u}]_{i,j} \equiv u_i u_j$.

1.2.2 Total energy balance

Up to this point, emphasis has been put on purely mechanical aspects of motion. However, common experience makes it plain that mechanical action does not always give rise to mechanical effects alone.

Taking the scalar product of (1.9) with $\mathbf{u}$ and using equation (1.5) we get the *mechanical energy equation*

$$\partial_t \left(\tfrac{1}{2} \varrho |\mathbf{u}|^2 \right) + \mathrm{div}_x \left(\tfrac{1}{2} \varrho |\mathbf{u}|^2 \mathbf{u} \right) - \mathrm{div}_x (\mathbb{T}\mathbf{u}) = -\mathbb{T} : \nabla_x \mathbf{u} + \varrho \mathbf{f} \cdot \mathbf{u} \qquad (1.10)$$

with the specific *kinetic energy* $\frac{1}{2}|\mathbf{u}|^2$. Here, the symbol $\mathbb{A}:\mathbb{B}$ stands for the *scalar product*,

$$\mathbb{A}:\mathbb{B} = \sum_{i,j=1}^{N} A_{i,j} B_{i,j}.$$

Equation (1.10) contains a non-conservative term $\mathbb{T}:\nabla_x \mathbf{u}$ which is responsible for changes of *internal energy e*. In accordance with the *first law of thermodynamics*, the inertial contribution (1.10) to the *total energy* of the body has to be matched by appropriate changes of its *internal energy* associated with restoring forces (e.g., with compressibility of the fluid), and energy dissipation into heat.

The *total energy*

$$\varrho\left(\tfrac{1}{2}|\mathbf{u}|^2 + e\right)$$

is a conserved quantity satisfying an integral identity

$$\frac{\mathrm{d}}{\mathrm{d}t}\int_{\mathbf{X}(t,B)} \varrho(t,\mathbf{x})\left(\tfrac{1}{2}|\mathbf{u}(t,\mathbf{x})|^2 + e(t,\mathbf{x})\right)\mathrm{d}\mathbf{x}$$

$$= \int_{\partial\mathbf{X}(t,B)} (\mathbb{T}(t,\mathbf{x})\mathbf{u}(t,\mathbf{x}) - \mathbf{q}(t,\mathbf{x}))\cdot\mathbf{n}\,\mathrm{d}\sigma$$

$$+ \int_{\mathbf{X}(t,B)} \varrho(t,\mathbf{x})\,\mathbf{f}(t,\mathbf{x})\cdot\mathbf{u}(t,\mathbf{x})\,\mathrm{d}\mathbf{x}, \tag{1.11}$$

where e denotes the *specific internal energy*, and $\mathbf{q}$ is the *energy flux* directly related to the transfer of heat.

Similarly, as above, one can use the convection theorem and Green's formula to deduce the *energy equation*:

$$\partial_t\left(\varrho(\tfrac{1}{2}|\mathbf{u}|^2 + e)\right) + \mathrm{div}_x\left(\varrho(\tfrac{1}{2}|\mathbf{u}|^2 + e)\mathbf{u}\right) + \mathrm{div}_x\mathbf{q} = \mathrm{div}_x(\mathbb{T}\mathbf{u}) + \varrho\mathbf{f}\cdot\mathbf{u}. \tag{1.12}$$

The *continuity equation* (1.5), the *momentum equation* (1.9), and the *energy equation* (1.12) represent the most general system governing the time evolution of a body in continuum mechanics.

The total energy balance (1.12) may be split into (1.10) and the *internal energy equation*:

$$\partial_t(\varrho e) + \mathrm{div}_x(\varrho e\mathbf{u}) + \mathrm{div}_x\mathbf{q} = \mathbb{T}:\nabla_x\mathbf{u}. \tag{1.13}$$

However, we must be careful to remember that the mechanical energy equation (1.10) was derived from the laws of mechanics, and that the total energy equation (1.12) and the internal energy equation (1.13) are two distinct equations with a proper physical meaning. In particular, it is worth noting that the dissipation function $\mathbb{T} : \nabla_x \mathbf{u}$, which appears in both (1.10) and (1.13) but *not* in (1.12), was derived from the differential form of Newton's second law expressed through (1.9), that is, under the assumption of smoothness of the motion. We will come back to this issue in Chapter 4 when a variational formulation of these equations will be discussed.

1.3 Constitutive equations

As already pointed out at the beginning of this chapter, the principal assumption adopted in this book is that the *state* of a body in motion at a given instant $t \in I$ is completely characterized by three macroscopic quantities: the density ϱ, the velocity $\mathbf{u}$, and the *temperature* ϑ. The physical properties of a particular material will be reflected through *constitutive equations* relating the state variables to other quantities appearing in the system (1.5), (1.9), (1.12)—the *stress tensor* $\mathbb{T}$, the *specific internal energy* e, and the *energy flux* $\mathbf{q}$.

A well accepted mathematical definition of a *fluid* reads as follows: when a shear stress is applied to any fluid, the fluid will deform continuously so long as the shear stress is active (see e.g. [61]). Equivalently, one can say that a fluid does not suppport shear stress when in equilibrium. Accordingly, the *stress tensor* $\mathbb{T}$ of a general fluid obeys *Stokes' law*

$$\mathbb{T} = \mathbb{S} - p\mathbb{I}, \tag{1.14}$$

where p is a scalar function termed *pressure*, and $\mathbb{S}$ denotes the *viscous stress tensor*, which characterizes the measure of resistance of the fluid to flow. As is to be expected, viscosity represents the mechanism by which the mechanical energy is transported into heat. The quantity $\mathbb{T} : \nabla_x \mathbf{u}$ appearing as an internal energy source on the right-hand side of (1.13) reads

$$\mathbb{T} : \nabla_x \mathbf{u} = \mathbb{S} : \nabla_x \mathbf{u} - p \operatorname{div}_x \mathbf{u},$$

where the former term, called the *dissipative function*, stands for a real (irreversible) *dissipation* of the mechanical energy into heat while the latter represents the energy change due to the work of compression.

1.3.1 Viscous dissipation

There are two sources of *viscous dissipation* in a fluid: (i) the *shear (or "genuine")* *viscosity* resulting in the departure of the tensor $\mathbb{S}$ from its isotropic form, (ii) the *bulk viscosity* related to irreversibility due to delays in attaining thermodynamic equilibrium. The bulk viscosity results from the molecular motion relative to the macroscopic velocity $\mathbf{u}$, and it is set at zero in the case of a monoatomic gas (see Chapter 1 of [80]). Accordingly, the viscous stress tensor $\mathbb{S}$ can be written as a sum of two orthogonal components

$$\mathbb{S} = \left(\mathbb{S} - \frac{1}{N} \text{ trace}[\mathbb{S}]\,\mathbb{I}\right) + \frac{1}{N} \text{ trace}[\mathbb{S}]\,\mathbb{I}, \qquad (1.15)$$

where the former represents the shear viscosity while the latter corresponds to the bulk viscosity. Here, orthogonal means with respect to the standard scalar product $\mathbb{A} : \mathbb{B}$ of tensors $\mathbb{A}$, $\mathbb{B}$.

In accordance with the principle of material frame indifference, the viscous stress tensor $\mathbb{S}$ must depend on the *symmetric part* $\mathbb{D}_x$ of the velocity gradient,

$$\mathbb{D}_x(\mathbf{u}) \equiv \tfrac{1}{2}(\nabla_x \mathbf{u} + \nabla_x \mathbf{u}^t),$$

its invariants, and possibly other scalar state variables like ϱ and ϑ. If the fluid is *isotropic* and

$$\mathbb{S} = \mathcal{F}(\mathbb{D}_x(\mathbf{u})),$$

then

$$\mathbb{O}\,\mathbb{S}\,\mathbb{O}^t = \mathcal{F}(\mathbb{O}\,\mathbb{D}_x(\mathbf{u})\,\mathbb{O}^t) \quad \text{for any unitary matrix } \mathbb{O} \in SO(N)$$

(see e.g. Chapter 2 of [117]).

An important class of fluids that occupies a central place in mathematical theory is represented by *linearly viscous* (Navier–Stokes, *Newtonian*) fluids for which the *viscous stress tensor* $\mathbb{S}$ depends linearly on the symmetric part of the velocity gradient $\mathbb{D}_x(\mathbf{u})$. In accordance with the general principles delineated above, the only admissible form of $\mathbb{S}$ reads

$$\mathbb{S} = 2\mu\,\mathbb{D}_x(\mathbf{u}) + \lambda\,\text{trace}[\mathbb{D}_x(\mathbf{u})]\,\mathbb{I}, \qquad (1.16)$$

where μ and λ are called *viscosity coefficients* (see Section 1.3 of Chapter 1 in [22]).

From the physical point of view, it is more natural to use the representation (1.15), which can be written as

$$\mathbb{S} = 2\mu\left(\mathbb{D}_x(\mathbf{u}) - \frac{1}{N}\text{div}_x\mathbf{u}\,\mathbb{I}\right) + \zeta\,\text{div}_x\mathbf{u}\,\mathbb{I},$$

where μ is the *shear viscosity coefficient* and $\zeta \equiv \lambda + (2/N)\mu$ the *bulk viscosity coefficient*. While μ should be positive for any "genuinely" viscous fluid, ζ may vanish, as is the case for a monoatomic gas.

The viscosity coefficients are quantities that may depend on the values of other state variables such as ϱ and ϑ. Experiments show that the viscosity of fluids is quite sensitive to changes in temperature ϑ. From this point of view, there is a major difference between *gases* and *liquids*: viscosity of gases increases with temperature, whereas the viscosity of liquids decreases.

It can be shown by methods of *statistical thermodynamics* that $\mu \approx c\sqrt{\vartheta}$ for a gas under normal conditions. At the same time, the coefficients of viscosity in gases show only little dependence on the density (see e.g. Chapter 10 in [20]).

For the purpose of this book, we shall assume that the fluid undergoes only mild changes of temperature during its evolution, and we shall consider both μ and λ constant. This is probably the most questionable point of the mathematical theory developed in what follows.

1.3.2 Pressure–density–temperature (pdt) state equation

In order to get a better understanding of the *constitutive equation* relating the *pressure* to the other state variables ϱ and ϑ, we have to abandon the macroscopic description and take into account the microscopic parameters describing the motion of individual molecules (atoms) of the fluid. As one would expect, there will be a substantial difference between *gases* and *liquids* due to the different relative distance of molecules (atoms) in both media.

Physically, a *perfect gas* is a medium in which at each instant only a very small portion of the molecules are sufficiently close to others. The pressure in a perfect gas is purely of thermal origin; it is related to the transfer of momentum by particles participating in the thermal motion, and it should be always proportional to the temperature. For a perfect gas, a very good approximation of p is provided by *Boyle's law*

$$p = p(\varrho, \vartheta) = \mathrm{R}\varrho\vartheta, \tag{1.17}$$

where R is a constant inversely proportional to the mean molecular weight of the gas.

A *liquid* can be described as a material that conforms to the shape of container, without necessarily filling its whole volume, when placed in it. The intermolecular forces in a liquid are no longer negligible due to a relatively small distance between individual molecules. There are *binding forces* acting on a relatively long distance as well as very strong *repulsive forces* due to intersection of electron clouds when the distance is extremely small (when ϱ is large). The presence of these phenomena results in a pressure contribution which is, in fact, independent of the temperature. As this pressure is the same for all temperatures up to absolute zero, it is sometimes called *cold pressure*. It is known that the pressure of highly condensed cold matter is proportional to $\varrho^{5/3}$ (see Chapters 3, 11 of [126]). The cold pressure need not be a monotone function of the density due to the presence of binding forces. A very elementary model capturing these phenomena was suggested by *van der Waals* (see Section 2.1 in [72]):

$$p = p(\varrho, \theta) = \frac{\mathrm{R}\varrho\vartheta}{1 - b\varrho} - a\varrho^2$$

with positive parameters a, b. Note that the equation becomes singular for $\varrho b = 1$ as the constant b is intended to correct for the finite volume occupied by the molecules.

In this book, we consider a general constitutive relationship

$$p = p(\varrho, \vartheta) = p_{\mathrm{e}}(\varrho) + p_{\mathrm{th}}(\varrho, \vartheta), \qquad (1.18)$$

where the *thermal pressure* p_{th} is a linear function of the temperature, that is,

$$p_{\mathrm{th}}(\varrho, \vartheta) = \vartheta \frac{\partial p}{\partial \vartheta}(\varrho) \equiv \vartheta p_\vartheta(\varrho). \qquad (1.19)$$

Moreover, we shall assume that p_ϑ is a non-decreasing function of the density vanishing for $\varrho = 0$.

The physical meaning of equation (1.18) is that the pressure can be regarded as being exerted by two mechanisms acting independently and additively: (i) the *elastic pressure* p_{e}, which is the same at a constant density for all values ϑ up to the absolute zero temperature, and which is evidently related to the part arising from attractive and repulsive intermolecular forces; (ii) the *thermal pressure* p_{th} attributed to the momentum transfer by the random translational motion of molecules (cf. Chapter 5 in [15]).

From the mathematical point of view, equation (1.18) can be understood as the first two terms in the Taylor expansion:

$$p(\varrho, \vartheta) = p(\varrho, \vartheta_0) + (\vartheta - \vartheta_0)\frac{\partial p}{\partial \vartheta}(\varrho, \vartheta_0) + \text{higher order terms}$$

for a given $\vartheta_0 > 0$.

The pressure in a real gas can be expressed by means of the so-called *virial series* in the form:

$$p(\varrho, \vartheta) \approx R\vartheta \sum_{i=1}^{\infty} B_i \varrho^i,$$

where $B_1 = 1$, and the other virial coefficients B_i, $i = 1, 2, \ldots$ are functions of ϑ (see e.g. Chapter 1 of [10]). One of the best approximations based on this approach is the *Beattie–Bridgman state equation* of the form:

$$p(\varrho, \vartheta) = R\vartheta\varrho + \beta\varrho^2 + \gamma\varrho^3 + \delta\varrho^4, \qquad (1.20)$$

where

$$\beta = B_0 R\vartheta - A_0 - cR\vartheta^{-2},$$

$$\gamma = -B_0 bR\vartheta + A_0 a - B_0 cR\vartheta^{-2},$$

$$\delta = B_0 bcR\vartheta^{-2}.$$

The product $B_0 b$ is usually extremely small when compared with other coefficients—one has $b \equiv 0$ for monoatomic gases like helium or argon, $B_0 b \approx 10^{-5}$ for nitrogen, oxygen, and air. On the other hand, changes of the coefficients proportional to ϑ^{-2} do not exceed 10% of the constant terms $A_0 a$ and A at temperatures close to room temperature (see Section 3.4 of Chapter 3 and Section 10.10 of Chapter 10 in [121]). Accordingly, a state equation of type (1.18), (1.19) is a very good approximation of (1.20).

The pressure in liquids can be expressed through (1.18) with the *elastic pressure* p_e proportional to $\varrho^{5/3}$ for large values of ϱ, and the *thermal pressure* p_{th} given by *Boyle's law* (1.17) (see Chapter 11 of [126]). The pressure in the so-called *Lennard–Jones–Devonshire model* of liquids can be approximated through

$$p(\varrho, \vartheta) = c_1 \vartheta \varrho + c_2(\varrho^5 - \varrho^3), \quad c_1, c_2 > 0,$$

which is of type (1.18) (see Section 2.2 of Chapter 2 of [72]).

Quite recently, there have been successful attempts to develop models of nuclear motion based on hydrodynamics. Wong ([123]) showed that such a theory can be coherently built from the time dependent *Hartree–Fock approximation*. The elastic pressure in a simplified model of a *nuclear fluid* can be expressed as

$$p_e = p^{(q)} + p^{(i)},$$

where

$$p^{(q)}(\varrho) = c_1 \varrho^{5/3}, \quad c_1 > 0,$$

is a quantum term calculated on the basis of the Thomas–Fermi–Weizsäcker approximation, and

$$p^{(i)} = c_2 \varrho^3 - c_3 \varrho^2, \quad c_2, c_3 > 0,$$

is related to the so-called Skyrme interaction. The thermal pressure p_{th} is given by Boyle's law (1.17) (see [35]).

1.3.3 The second law of thermodynamics, entropy

Given the (pdt) state equation discussed in the preceding section, the specific *internal energy* $e = e(\varrho, \vartheta)$ must satisfy *Maxwell's relationship*

$$\frac{\partial e}{\partial \varrho} = 1/\varrho^2 (p(\varrho, \vartheta) - \vartheta p_\vartheta(\varrho)) = \frac{p_e(\varrho)}{\varrho^2}$$

(see e.g. Chapter 3 in [9]).

Accordingly, e can be written in the form:

$$e(\varrho, \vartheta) = P_e(\varrho) + Q(\vartheta), \qquad (1.21)$$

with the *elastic potential* P_e,

$$P_e(\varrho) \equiv \int_1^\varrho \frac{p_e(z)}{z^2}\, dz, \qquad (1.22)$$

and the *thermal energy* contribution Q depending solely on the temperature. Here,

$$c_v(\vartheta) \equiv \frac{\partial e}{\partial \vartheta} = \frac{\partial Q}{\partial \vartheta}$$

represents the *specific heat at constant volume* (heat input per unit mass per unit increase in temperature). Under normal temperature changes, it is customary to take c_v constant. In general, we have

$$Q(\vartheta) \equiv \int_0^\vartheta c_v(z)\, dz.$$

The elastic pressure potential P_e satisfies the following differential equation:

$$\partial_t(\varrho P_e(\varrho)) + \operatorname{div}_x(\varrho P_e(\varrho)\mathbf{u}) + p_e(\varrho)\operatorname{div}_x \mathbf{u} = 0;$$

which is easy to deduce multiplying the continuity equation (1.5) by the derivative $(\varrho P_e(\varrho))'$. Note that the function P_e satisfies the identity

$$\varrho P'_e(\varrho) - P_e(\varrho) = p_e(\varrho)$$

which is a non-homogeneous Euler equation.

Consequently, equation (1.13) can be written as the *thermal energy equation*

$$\partial_t(\varrho Q(\vartheta)) + \operatorname{div}_x(\varrho Q(\vartheta)\mathbf{u}) + \operatorname{div}_x \mathbf{q} = \mathbb{S} : \nabla_x \mathbf{u} - \vartheta p_\vartheta \operatorname{div}_x \mathbf{u}. \qquad (1.23)$$

Introducing the *specific entropy* $s = s(\varrho, \vartheta)$ through the thermodynamics relationship

$$\frac{\partial s}{\partial \varrho} = \frac{1}{\vartheta}\left(\frac{\partial e}{\partial \varrho} - \frac{p}{\varrho^2}\right), \qquad \frac{\partial s}{\partial \vartheta} = \frac{1}{\vartheta}\frac{\partial e}{\partial \vartheta},$$

that is,

$$s = s(\varrho, \vartheta) \equiv \int_1^\vartheta \frac{c_v(z)}{z}\, dz - P_\vartheta(\varrho) \qquad (1.24)$$

with the *thermal pressure potential* P_ϑ,

$$P_\vartheta(\varrho) \equiv \int_1^\varrho \frac{p_\vartheta(z)}{z^2}\, dz, \qquad (1.25)$$

we easily deduce from (1.5), (1.23) the *entropy equation*

$$\partial_t(\varrho s) + \mathrm{div}_x(\varrho s \mathbf{u}) + \mathrm{div}_x\left(\frac{\mathbf{q}}{\vartheta}\right) = \frac{\mathbb{S} : \nabla_x \mathbf{u}}{\vartheta} - \frac{\mathbf{q} \cdot \nabla_x \vartheta}{\vartheta^2} \tag{1.26}$$

(see e.g. Chapter 3 in [9]).

A motion, together with an entropy, constitutes a thermodynamics process of the body. The second law of thermodynamics requires the *Clausius–Duhem inequality*

$$\frac{\mathrm{d}}{\mathrm{d}t} \int_{\mathbf{X}(t,B)} \varrho(t,\mathbf{x}) s(t,\mathbf{x}) \, \mathrm{d}\mathbf{x} + \int_{\partial \mathbf{X}(t,B)} \frac{\mathbf{q}(t,\mathbf{x})}{\vartheta(t,\mathbf{x})} \, \mathrm{d}\sigma \geq 0$$

to be satisfied for any domain $B \subset \Omega$ (see e.g. Lecture 27 in [115]).

In the context of smooth processes, this relation may be regarded as a condition of material response which induces restrictions on constitutive relations. Accordingly, the right-hand side of (1.26) must be non-negative; therefore

$$\mathbb{S} : \nabla_x \mathbf{u} \geq 0, \quad \mathbf{q} \cdot \nabla_x \vartheta \leq 0.$$

In particular, the *viscosity coefficients* μ and ζ for a *Newtonian fluid* introduced in (1.16) should satisfy

$$\mu \geq 0, \quad \zeta \equiv \lambda + \frac{2}{N}\mu \geq 0. \tag{1.27}$$

Thus the bilinear form

$$\mathbb{S}(\mathbf{u}) : \nabla_x \mathbf{v} \equiv \mathbb{S}(\mathbb{D}_x(\mathbf{u})) : \mathbb{D}_x(\mathbf{v})$$

is positively definite on the space of symmetric tensors on R^N provided $\mu > 0$.

By the same token, the scalar product of the internal energy flux $\mathbf{q}$ with $\nabla_x \vartheta$ should be negative. The simplest case is expressed through *Fourier's law*:

$$\mathbf{q} = -\kappa(\vartheta)\nabla_x \vartheta, \quad \kappa \geq 0, \tag{1.28}$$

where κ is the *heat conductivity coefficient* which may depend on the temperature.

For *perfect gases* with constant specific heat, the statistical thermodynamics predicts a constant ratio of the viscosity coefficient μ to κ called the *Prandtl number*:

$$Pr \equiv c(N)\frac{\mu}{\kappa}.$$

At high temperatures, however, a completely different mechanism of heat energy transfer appears due to *radiation*. As a result, the heat conductivity coefficient κ

becomes a rather sensitive function of temperature. The speed of propagation by radiation is much higher than by pure molecular diffusion, which corresponds to

$$\kappa \approx \vartheta^{\alpha} \quad \text{for } \alpha \approx 4.5 - 5.5$$

for large values of ϑ (cf. Section 2 of Chapter 10 in [126]).

1.4 Barotropic flows

A fluid flow is said to be in a *barotropic regime* when the pressure p depends solely on the density. At a purely formal level, this can be viewed as taking $p_{\text{th}} \equiv 0$ in (1.18). However, this should never be understood in the sense that the only source of pressure in the fluid results from intermolecular forces—a hypothesis that is physically unsustainable at least for fluids at normal temperature. The physical background can be rather different for problems yielding the same system of mathematical equations. The common feature for all barotropic flows is that the *specific internal energy* is now interpreted as

$$e \equiv P(\varrho), \quad \text{with } P(\varrho) \equiv \int_{1}^{\varrho} \frac{p(z)}{z^2} \, dz;$$

the continuity equation (1.5), and the momentum equations (1.9) together form a closed system; and the energy equation (1.12) is replaced (rather inconsistently) by a relation

$$\partial_t \left(\varrho \left(\tfrac{1}{2} |\mathbf{u}|^2 + P(\varrho) \right) \right) + \operatorname{div}_x \left(\varrho \left(\tfrac{1}{2} |\mathbf{u}|^2 + P(\varrho) \right) \mathbf{u} + p\mathbf{u} \right)$$
$$= \operatorname{div}_x (\mathbb{S}\,\mathbf{u}) - \mathbb{S} : \nabla_x \mathbf{u} + \varrho \mathbf{f} \cdot \mathbf{u} \tag{1.29}$$

which can be directly deduced from (1.5), (1.9). In other words, the changes of internal energy in a *barotropic flow* can be interpreted as purely mechanical ones resulting from work done on a volume element of the fluid. In the remaining part of this section, we give three rather different examples of barotropic flows.

1.4.1 Isothermal flows

The simplest example of a barotropic flow is an *isothermal flow* where the temperature ϑ is assumed to be constant. If the medium is a perfect gas, the pressure p is given, in accordance with *Boyle's law*, by the constitutive equation

$$p(\varrho) = R\vartheta_0\varrho \quad \text{with positive constants } R, \vartheta_0.$$

Another example of an "isothermal flow" is a completely degenerate electron gas at a temperature close to absolute zero. The pressure is then given by a *barotropic constitutive equation* which is the same as that of a monoatomic gas, namely,

$$p(\varrho) = a\varrho^{5/3}, \quad a > 0$$

(see e.g. Chapter 3 of [126]). The same equation of state also describes a non-relativistic completely degenerate neutron gas which constitutes the matter of neutron stars.

Further examples may be obtained, of course, taking $\vartheta = \vartheta_0$ in a virial series determining the pressure for a given fluid.

1.4.2 Isentropic flows

If both conduction of heat and its generation by dissipation of mechanical energy can be neglected, the right-hand side of (1.26) vanishes leaving the entropy constant along streamlines. If this is the case, the temperature ϑ is uniquely determined as a function of ϱ through (1.24) yielding a barotropic state equation for the pressure. If the fluid is a perfect gas obeying Boyle's law (1.17), we get

$$p = p(\varrho) = a\varrho^{\gamma}, \qquad a > 0, \quad \gamma = \frac{R + c_v}{c_v}, \qquad (1.30)$$

where $\gamma > 1$ is termed the *adiabatic constant*. Typical values of γ range from a maximum $\frac{5}{3}$ for *monoatomic gases*, through $\frac{7}{5}$ for *diatomic gases* including air, to lower values close to 1 for *polyatomic gases* at high temperature. In general, we have

$$\gamma = \frac{N + m + 2}{N + m},$$

where m is the number of non-translational degrees of freedom of the molecules of the fluid.

Similarly as in the case of an isothermal flow, more examples can be obtained by setting s constant in (1.24) and substituting the corresponding expression for the temperature in the thermal pressure p_{th}.

1.4.3 Barotropic flows in nuclear astrophysics

This last and rather exotic example of a *barotropic flow* was discussed in [36]. Using the finite temperature *Hartree–Fock theory* one can derive the following constitutive equation for the pressure in a *nuclear fluid*:

$$p_G(\varrho, \vartheta) = a(1 + \sigma)\varrho^{2+\sigma} - b\varrho^2 + k\vartheta \sum_{n=1}^{\infty} B_n \varrho^n, \qquad (1.31)$$

where k is the Boltzmann constant, and where the last series converges very rapidly because of the fast decay of the sequence B_n.

Consider a realistic situation when a photon assembly is superimposed on the nuclear matter background. If this radiation is in quasi-local thermodynamic equilibrium with the nuclear fluid, one can show that the resulting mixture nucleons–photons can be described by a one-fluid Navier–Stokes system provided one adds to (1.31) a Stefan–Boltzmann contribution of black-body type:

$$p_R(\vartheta) = d\vartheta^4, \quad d > 0. \tag{1.32}$$

Now a further simplification can be introduced leading to the so-called *Eddington standard model*. This approximation means to assume that the ratio of the total pressure $p = p_G + p_R$ and the radiative pressure p_R is a pure constant. Although very crude, this model is in a very good agreement with more sophisticated ones, in particular, for the sun. Supposing $\sigma = 1$ and keeping only the lowest order term in the series in (1.31), one obtains

$$2a\varrho^3 - b\varrho^2 + kB_1\vartheta\varrho = \nu\vartheta^4 \quad \text{for a certain constant } \nu.$$

Solving this algebraic equation to leading order (high temperature) one gets

$$\vartheta \approx \text{const } \varrho^{3/4}$$

which gives rise to the *pressure law*

$$p(\varrho) = a_1\varrho^3 - b\varrho^2 + a_2\varrho^{7/4}, \quad a_1, a_2 > 0. \tag{1.33}$$

1.5 The Navier–Stokes system

In accordance with our previous discussion, we shall assume that the *evolution* of a fluid occupying a domain $\Omega \subset R^N$ during a time interval $I \subset R$ is completely described by the values of the three *macroscopic state variables*—the *density* $\varrho = \varrho(t, \mathbf{x})$, the *velocity* $\mathbf{u} = \mathbf{u}(t, \mathbf{x})$, and the *temperature* $\vartheta = \vartheta(t, \mathbf{x})$ obeying the *Navier–Stokes (N–S) system* of equations:

---

equation of continuity:

$$\partial_t \varrho + \text{div}_x(\varrho \mathbf{u}) = 0; \tag{1.34}$$

momentum equation:

$$\partial_t(\varrho \mathbf{u}) + \text{div}_x(\varrho \mathbf{u} \otimes \mathbf{u}) + \nabla_x p = \text{div}_x \mathbb{S} + \varrho \mathbf{f}; \tag{1.35}$$

thermal energy equation:

$$\partial_t(\varrho Q(\vartheta)) + \text{div}_x(\varrho Q(\vartheta)\mathbf{u}) + \text{div}_x \mathbf{q} = \mathbb{S} : \nabla_x \mathbf{u} - \vartheta p_\vartheta \text{div}_x \mathbf{u}; \tag{1.36}$$

---

in the domain $I \times \Omega$, where $Q \in C^2[0, \infty)$ is a given function such that

$$Q(\vartheta) \equiv \int_0^\vartheta c_v(z)\,\mathrm{d}z, \quad c_v(z) \geq c_v > 0 \text{ for all } z \geq 0.$$

The quantities p, $\mathbb{S}$, and $\mathbf{q}$ are determined in terms of ϱ, $\nabla_x \mathbf{u}$, and ϑ through a family of *constitutive equations*:

Newton's law of viscosity:

$$\mathbb{S} = \mu(\nabla_x \mathbf{u} + \nabla_x \mathbf{u}^t) + \lambda \operatorname{div}_x \mathbf{u}\,\mathbb{I}, \tag{1.37}$$

with constant viscosity coefficients μ, λ,

$$\mu > 0, \quad \lambda + \frac{2}{N}\mu \geq 0 \tag{1.38}$$

the (pdt) state equation:

$$p = p_{\mathrm{e}}(\varrho) + \vartheta p_\vartheta(\varrho) \tag{1.39}$$

Fourier's law:

$$\mathbf{q} = -\kappa(\vartheta)\nabla_x \vartheta, \quad \kappa > 0 \tag{1.40}$$

The boundary $\partial\Omega$ is considered as a *kinematic boundary*; that means a surface that separates permanently the fluid from "rest of the world". To fix ideas, we suppose that $\Omega \subset R^N$ is a bounded domain with $\partial\Omega$ of class $C^{2+\nu}$ for a certain $\nu > 0$; in particular, the outer normal vector $\mathbf{n}$ is well defined. Moreover, the system (1.34)–(1.36) must be supplemented by a family of *boundary conditions* in order to obtain a mathematically well-posed problem. Viscous fluids are known to adhere to the kinematic boundary as confirmed by numerous experiments. Accordingly, we shall assume

the no-slip boundary condition:

$$\mathbf{u}(t, \mathbf{x}) = 0 \quad \text{for all } t \in I, \ \mathbf{x} \in \partial\Omega \tag{1.41}$$

conservative boundary condition:

$$\mathbf{q}(t, \mathbf{x}) \cdot \mathbf{n}(\mathbf{x}) = 0 \quad \text{for all } t \in I, \ \mathbf{x} \in \partial\Omega. \tag{1.42}$$

Given a reference time $t_I \in \overline{I}$ the *initial state* of the fluid will be given through

the initial conditions:

$$\varrho(t_I, \mathbf{x}) = \varrho_0(\mathbf{x}),$$
$$(\varrho \mathbf{u})(t_I, \mathbf{x}) = \mathbf{m}_0(\mathbf{x}),$$
$$(\varrho Q(\vartheta))(t_I, \mathbf{x}) = \chi_0(\mathbf{x}) \quad \text{for all } \mathbf{x} \in \Omega. \tag{1.43}$$

Although the above system of equations was derived for a *non-dilute fluid*, that is, for strictly positive densities, it will be convenient to consider the density ϱ_0—a non-negative function which may vanish on some part of Ω. Accordingly, the initial data must satisfy a *compatibility condition*

$$\mathbf{m}_0 = 0, \quad \chi_0 = 0 \text{ on the set} \{\mathbf{x} \in \Omega \,|\, \varrho_0(\mathbf{x}) = 0\}. \tag{1.44}$$

Physically, the regions where the density vanishes correspond to possible vacuum zones that may appear. This problem will be discussed in Chapter 4.

As a matter of fact, it seems more convenient to work with the initial condition stated in terms of the temperature itself. One can assume

$$\varrho Q(\vartheta)(t_I, \mathbf{x}) = \varrho_0(\mathbf{x}) Q(\vartheta_0)(\mathbf{x}),$$

with a given function ϑ_0. For obvious physical reasons, we shall always assume ϑ_0 strictly positive on Ω even though such a stipulation does not have any meaning on the set

$$\{\varrho_0 = 0\} \equiv \{\mathbf{x} \in \Omega \,|\, \varrho_0(\mathbf{x}) = 0\}.$$

Finally, we shall assume that the *total energy* of the system at the time t_I is finite, which gives rise, with the convention (1.44), to the hypothesis

$$E_0 = \int_\Omega \frac{1}{2} \frac{|\mathbf{m}_0(\mathbf{x})|^2}{\varrho_0(\mathbf{x})} + \varrho_0(\mathbf{x}) P_e(\varrho_0(\mathbf{x})) + \chi_0(\mathbf{x}) \, d\mathbf{x} < \infty. \tag{1.45}$$

1.6 Bibliographical notes

1.1–1.2 The material collected in these sections is classical. The concept of bodies, configurations and motions follows the presentation by Truesdell [115, 116] and Truesdell and Rajagopal [117]. The reader interested in a more "axiomatic" approach can consult the monograph by Gallavotti [59]. The classical reference materials are, of course, the works of Lamb [76], Batchelor [9], or Landau and Lifschitz [77].

1.3–1.4   Constitutive equations are crucial for a consistent mathematical theory to be developed. They determine the function spaces framework as well as the functional analytic methods to be used to attack the problem.

There is a nice exposition on linearly viscous fluids done by Chorin and Marsden [22]. Here, one can find the basic concepts including a rigorous derivation of the specific form of the viscous stress tensor and the interpretation of the viscosity coefficients. More information on viscous dissipation is available in [80].

It is well known from experimental results that there exist stronger dissipative mechanisms in nature than those captured by the classical linear Stokes' law. In the linear theory of *multipolar fluids*, the constitutive laws, in particular the stress tensor $\mathbb{S}$, depends not only on the first spatial gradients of the velocity field $\mathbf{u}$ but also on higher order gradients up to order $2k - 1$ for so-called k-polar fluids. In the work of Nečas and Šilhavý [100], an axiomatic theory of viscous multipolar fluids is developed on the basis of the theory of elastic materials due to Green and Rivlin [62]. In this theory, the *viscous stress tensor* $\mathbb{S}$ takes the form:

$$\mathbb{S} = \sum_{j=0}^{k-1}((-1)^j\mu_j\Delta^j\mathbb{D}_x(\mathbf{u}) + \lambda_j\Delta^j\mathrm{div}_x\mathbf{u}\mathbb{I}) + \omega(\mathbb{D}_x(\mathbf{u})) + \beta\,\mathrm{div}_x\mathbf{u}\,\mathbb{I},$$

where β and ω may depend on the invariants $|\nabla_x\mathbf{u}|$, $\mathrm{div}_x\mathbf{u}$, $\det\nabla_x\mathbf{u}$.

The existence of so-called measure-valued solutions for this problem was proved in [93]. More information can be found in [98], [99] and [90], [91].

The material concerning the (pdt) state equation is mostly taken from the textbook by Van Wyllen and Sonntag [121]. The constitutive equations for liquids containing so-called cold pressure are discussed in Chapter 11 of [126]. Here, one can also find a more exact equation for the pressure taking into account the so-called electronic pressure:

$$p = p(\varrho, \vartheta) = p_{\mathrm{e}}(\varrho) + \mathrm{R}\varrho\vartheta + \sqrt{\varrho}\vartheta^2.$$

The dependence on ϑ of the heat conductivity coefficient at high temperatures is discussed in Chapter 10 of [126]. A more recent approach to the relation between μ and κ based on the investigation of various collision operators in the Boltzmann equation can be found in [26].

As for the examples borowed from condensed matter physics as well as the general pressure density relations discussed in Section 1.3, we followed Chapters 1 and 11 of [126] and Chapter 2 of [72]. The non-monotone pressure density relations appear in simple models of phase transitions studied by Fan and Slemrod [43], where an extensive list of references on the subject can be found. Hydrodynamical models of cold nuclear matter were developed by Ducomet [34], Tang and Wong [110], and Wong [123]. Further examples can be found in [79], [96], or [70].

1.5 A detailed treatment of the less standard *no-stick boundary conditions* is done by Ebin [38]. More general examples including some *unilateral boundary conditions* are considered in the monograph of Antontsev et al. [5]. One can also study the problem posed on an unbounded spatial domain. This and other types of boundary conditions are examined in the monograph [85].

2

MATHEMATICAL PRELIMINARIES

The Navier–Stokes system introduced in Chapter 1 is a nonlinear problem. There are two types of nonlinearities: (i) *geometric nonlinearities* arising because of our choice of the Eulerian description, (ii) *physical nonlinearities* expressed through various constitutive equations. While the former type results from our choice of description of motion, the latter represents intrinsic properties of the material we deal with.

The mathematical theory developed in this book is based on the concept of *variational solutions* This type of solution was introduced by Leray in the context of the incompressible Navier–Stokes system (see [78]). In the variational (weak) formulation, the derivatives are understood in the sense of distributions, the equations being replaced by a family of integral identities which, in the case of the Navier–Stokes system, are much closer to the integral relations (1.4), (1.8), and (1.11) rather than to the corresponding partial differential euqations (1.5), (1.9), and (1.12). Variational solutions need not be differentiable nor even continuous. Pointwise values of the state variables are replaced by their integral means which, by the way, seems to be much closer to the concept of a "measurable quantity" in the physical sense.

Unlike linear equations, which are usually well posed on a large scale of functions spaces depending only on the regularity hypotheses imposed on the data, nonlinear problems are usually solvable only in the function spaces, the topology of which is determined by *a priori estimates*

In this chapter, we introduce and review some basic properties of the function spaces that are related to our problem. Then, in Chapter 3, we establish the fundamental *a priori* estimates that form the mathematical background of the theory of variational solutions developed later in this book.

2.1 Function spaces

In what follows, the symbol R^M denotes the M-dimensional Euclidean space. This space has a topological structure given by the standard scalar product and a measure structure induced by the Lebesgue measure $d\mathbf{y}$.

2.1.1 *Spaces of continuous and continuously differentiable functions*

For $O \subset R^M$ a bounded open set, X a topological metric space, we introduce the Fréchet space $C(\overline{O}; X)$ of continuous functions $v : \overline{O} \mapsto X$ endowed with a metric

$$d_{C(\overline{O};X)}[v,w] = \sup_{\mathbf{y}\in\overline{O}} d_X[v(\mathbf{y}), w(\mathbf{y})],$$

where d_X is the metric on X. If X is a Banach space, then $C(\overline{O};X)$ is also a Banach space with a norm

$$\|v\|_{C(\overline{O};X)} = \sup_{\mathbf{y}\in\overline{O}} \|v(\mathbf{y})\|_X.$$

The space $C(\overline{O};R)$ will be denoted as $C(\overline{O})$.

We report the following abstract version of the *Arzelà–Ascoli theorem*.

Theorem 2.1 [69, Chapter 7, Theorem 17] *Let $\overline{O} \subset R^M$ be compact and X a compact topological metric space endowed with a metric d_X. Let $\{v_n\}_{n=1}^{\infty}$ be a sequence of functions in $C(\overline{O};X)$ which is equi-continuous, that is, for any $\varepsilon > 0$ there is $\delta > 0$ such that*

$$d_X[v_n(\mathbf{y}), v_n(\mathbf{z})] \le \varepsilon \ \text{provided} \ |\mathbf{y} - \mathbf{z}| < \delta \ \text{independently of} \ n = 1, 2, \dots$$

Then $\{v_n\}_{n=1}^{\infty}$ is precompact in $C(\overline{O};X)$, that is, there exists a subsequence (not relabeled) and a function $v \in C(\overline{O};X)$ such that

$$\sup_{\mathbf{y}\in\overline{O}} d_X[v_n(\mathbf{y}), v(\mathbf{y})] \to 0 \ \text{as} \ n \to \infty.$$

Analogously, as for functions ranging in a finite-dimensional space, we can define a partial derivative with respect to the variable y_i of $v : O \mapsto X$ as

$$\partial_{y_i} v(\mathbf{y}) \equiv \lim_{h\to 0} \frac{v(y_1, \dots, y_i + h, \dots, y_M) - v(y_1, \dots, y_M)}{h}$$

provided X is a Banach space and the limit exists with respect to the norm-topology on X.

Let k be a positive integer, and O a bounded open set in R^M. The space $C^k(\overline{O};X)$ consists of all functions $v : O \to X$ having all partial derivatives up to order k which can be extended as continuous functions on $\overline{O}$. Similarly, we introduce the set $C^{\infty}(\overline{O};X)$

$$C^{\infty}(\overline{O};X) \equiv \cap_{k\ge 0} C^k(\overline{O};X).$$

Moreover, the symbol

$$C^{k,\nu}(\overline{O};X), \quad k \text{ a non-negative integer}, \ \nu \in (0,1)$$

will denote the space of functions whose derivatives up to order k are ν-Hölder continuous on $\overline{O}$.

The spaces $C^k(O;X)$, where $O \subset R^M$ is an open set, consist of functions belonging to $C^k(K;X)$ for any compact $K \subset O$. Moreover, we introduce the space $\mathcal{D}(O, X)$ of smooth compactly supported functions:

$$\mathcal{D}(O; X) \equiv \{v \in C^\infty(O; X) \mid \operatorname{supp}[v] \text{ is compact in } O\};$$

which plays a crucial role in the theory of *generalized derivatives* (distributions) introduced below. Similarly to above, we write $\mathcal{D}(O) = \mathcal{D}(O; R)$.

Finally, the spaces $C_0^k(\overline{O}; X)$ are defined as the completion of $\mathcal{D}(O; X)$ with respect to the C^k-norm. Equivalently, if the open set O is sufficiently regular, one can say that $C_0^k(\overline{O}; X)$ consists of those functions from $C^k(\overline{O}; X)$ whose partial derivatives up to order k vanish on the boundary ∂O.

2.1.2 Spaces of integrable functions

A priori estimates for solutions of evolutionary problems in continuum mechanics are mostly derived from various kinds of energy or entropy balance equations. Accordingly, the bounds are given in terms of integral identities and, consequently, the most natural functional analytic framework is based on the Lebesgue spaces of integrable functions. One should always keep in mind that these are spaces formed by equivalent classes of functions which coincide up to a set of (Lebesgue) measure zero.

Let O be a measurable subset of the Euclidean space R^M, $M \geq 1$, endowed with the standard Lebesgue measure $\mathrm{d}y$, and let X be a Banach space. The *Lebesgue spaces* $L^p(O; X)$, $p \in [1, \infty]$ are defined as spaces of (classes of) Bochner measurable functions $v : O \to X$ with finite norm

$$\|v\|_{L^p(O;X)} \equiv \left(\int_O \|v\|_X^p \, \mathrm{d}y \right)^{1/p}, \quad p \in [1, \infty), \quad \|v\|_{L^\infty(O)} \equiv \operatorname*{ess\,sup}_{y \in O} \|v(y)\|_X.$$

Similarly to the above, we shall write $L^p(O) \equiv L^p(O; R)$.

The scalar valued functions in $L^p(O)$ satisfy *Hölder's inequality* (see e.g. [1, Theorem 2.3]):

$$\|v_1 v_2 \ldots v_n\|_{L^q(O)} \leq \|v_1\|_{L^{p_1}(O)} \|v_2\|_{L^{p_2}(O)} \cdots \|v_n\|_{L^{p_n}(O)}, \tag{2.1}$$

where

$$\frac{1}{q} = \sum_{i=1}^n \frac{1}{p_i}, \quad p_i, q \in [1, \infty], \quad i = 1, \ldots, n.$$

In the particular case $p_1 = p_2 = 2$, one recovers the *Cauchy–Schwartz inequality*. Finally, as a straightforward consequence of (2.1), we get the *interpolation inequality*

$$\|v\|_{L^q(O)} \leq \|v\|_{L^p(O)}^\alpha \|v\|_{L^r(O)}^{1-\alpha} \quad \text{with } \frac{1}{q} = \frac{\alpha}{p} + \frac{1-\alpha}{r}. \tag{2.2}$$

There is another way to introduce the Lebesgue spaces, namely to consider the completion of the space $\mathcal{D}(O; X)$ of smooth compactly supported functions with respect to the norm $\| \cdot \|_{L^p(O;X)}$. To this end, we introduce the standard *smoothing operators* $v \mapsto [v]_y^\varepsilon$ based on convolution with a sequence of regularizing kernels:

$$\theta_y^\varepsilon(\mathbf{y}) \equiv \frac{1}{\varepsilon^M} \theta\left(\frac{|\mathbf{y}|}{\varepsilon}\right) \quad \text{for } \mathbf{y} \in R^M, \ \varepsilon > 0, \tag{2.3}$$

where

$$\theta \in \mathcal{D}(-1,1), \quad \theta(-z) = \theta(z), \quad \int_{R^M} \theta(|z|)\, dz = 1, \quad \theta \text{ non-increasing on } [0,\infty).$$

For a function $v \in L^1(O;X)$, we set

$$[v]_y^\varepsilon(\mathbf{y}) \equiv (\theta_y^\varepsilon * v)(\mathbf{y}) \equiv \int_{R^M} \theta_y^\varepsilon(\mathbf{y} - \mathbf{z}) v(\mathbf{z})\, dz, \tag{2.4}$$

where v has been extended to be zero outside O. We have

$$[v]_y^\varepsilon \in \mathcal{D}(U_\varepsilon(O); X)$$

where $U_\varepsilon(O)$ is the ε-neighborhood of the set O.

The smoothing operators are bounded on $L^p(O;X)$ which can be deduced easily from Young's inequality:

$$\|[v]_y^\varepsilon\|_{L^p(O;X)} \leq \|v\|_{L^p(O;X)} \quad \text{for any } 1 \leq p \leq \infty \tag{2.5}$$

provided the norm on the right-hand side is finite.

Moreover, it is easy to see that

$$[v]_y^\varepsilon \to v \text{ in } C(R^M; X) \quad \text{as } \varepsilon \to 0$$

for any $v \in C(R^M; X)$ vanishing for $|\mathbf{y}| \to \infty$. Consequently, by virtue of the Banach–Steinhaus theorem,

$$[v]_y^\varepsilon \to v \text{ in } L^p(O;X) \quad \text{as } \varepsilon \to 0 \text{ for } 1 \leq p < \infty \tag{2.6}$$

provided $v \in L^p(O;X)$.

In particular, we have shown the following result.

Theorem 2.2   *Let $O \subset R^M$ be an open set and X a Banach space.*

Then the set $\mathcal{D}(O;X)$ of smooth compactly supported functions is dense in $L^p(O;X)$ for any $1 \leq p < \infty$. Moreover, if X is separable, then $L^p(O;X)$, $1 \leq p < \infty$, is separable.

To conclude, we recall a characterization of the dual space $[L^p(O;X)]^*$.

Theorem 2.3 [57, Theorem 1.14]   *Let $O \subset R^M$ be a measurable set, X a separable reflexive Banach space, and $1 \leq p < \infty$.*

Then any continuous linear form $\xi \in [L^p(O;X)]^*$ admits a unique representation $w_\xi \in L^{p'}(O;X^*)$,

$$\langle \xi, v \rangle = \int_O \langle w_\xi(\mathbf{y}), v(\mathbf{y}) \rangle \mathrm{d}\mathbf{y} \text{ for all } v \in L^p(O;X),$$

where

$$\frac{1}{p} + \frac{1}{p'} = 1.$$

Accordingly, the spaces $L^p(O;X)$ are reflexive for $1 < p < \infty$.

Remark The representation formula derived in Theorem 2.3 is due to Riesz. As a matter of fact, it is still valid if X is *either* reflexive *or* separable (see e.g. [39]).

Finally, we introduce the space $L^p_{\mathrm{loc}}(O;X)$ as the set of all functions $v : O \mapsto X$ belonging to $L^p(K;X)$ for any compact set $K \subset O$.

2.1.3 *Weak derivatives, Sobolev spaces*

The fundamental idea of the theory of *generalized derivatives* (distributions) is to identify a function v defined on an open set $\Omega \subset R^N$ with the family of its *integral averages*

$$v \approx \int_\Omega v(\mathbf{x}) \, \varphi(\mathbf{x}) \, \mathrm{d}\mathbf{x} \quad \text{for } \varphi \in \mathcal{D}(\Omega).$$

Note that this makes sense for any locally integrable function v.

More specifically, the space of *distributions* $\mathcal{D}'(\Omega)$ is defined as the set of all linear forms ξ on $\mathcal{D}(\Omega)$ which are continuous in the following sense:

$$\langle \xi, \eta_n \rangle \to \langle \xi, \eta \rangle \quad \text{as } n \to \infty$$

for any sequence $\{\eta_n\}_{n=1}^\infty \subset \mathcal{D}(\Omega)$ such that

(1) the support of η_n is contained in a fixed compact set $K \subset \Omega$ for all $n = 1, 2, \ldots$; and

(2) $\eta_n \to \eta$ in $C^k(\Omega)$ for any $k = 0, 1, \ldots$.

It is important that the set $\mathcal{D}(\Omega)$ itself can be identified in a *unique way* as a subspace of $\mathcal{D}'(\Omega)$ through a representation formula

$$v \mapsto \xi_v, \quad \langle \xi_v, \eta \rangle = \int_\Omega v \, \eta \mathrm{d}\mathbf{x} \quad \text{for all } \eta \in \mathcal{D}(\Omega). \tag{2.7}$$

Accordingly, any locally integrable function v admits a unique representation in $\mathcal{D}'(\Omega)$ via (2.7).

Now, it is natural to define a linear differential operator

$$\mathcal{L} = \sum_{i=1}^{m} a_i \partial_{\mathbf{x}}^{\mathbf{k}_i}, \quad a_i \in R$$

on $\mathcal{D}'(\Omega)$ through the integral formula

$$\langle \mathcal{L}\,\xi, \eta \rangle = \sum_{i=1}^{m} \int_{\Omega} (-1)^{|\mathbf{k}_i|}\, \xi\, \partial_{\mathbf{x}}^{\mathbf{k}_i} \eta\, \mathrm{d}\mathbf{x} \quad \text{for all } \eta \in \mathcal{D}(\Omega).$$

We shall say $\mathcal{L}v = w$, where w is a locally integrable function, if $\mathcal{L}v$ in $\mathcal{D}'(\Omega)$ can be indentified with w in the sense of (2.7).

Similarly, we can say that the inequality

$$\sum_{i=1}^{m} a_i \partial_{\mathbf{x}}^{\mathbf{k}_i}\, v \geq w \quad \text{holds in } \mathcal{D}'(\Omega)$$

for locally integrable functions v, w if and only if

$$\sum_{i=1}^{m} \int_{\Omega} (-1)^{|\mathbf{k}_i|} v\, \partial_{\mathbf{x}}^{\mathbf{k}_i} \eta\, \mathrm{d}\mathbf{x} \geq \int_{\Omega} w\eta\, \mathrm{d}\mathbf{x} \quad \text{for any } \eta \in \mathcal{D}(\Omega),\ \eta \geq 0.$$

Any distribution $\xi \in \mathcal{D}'(\Omega)$ can be regularized by the convolution operator $\xi \mapsto [\xi]_x^\varepsilon$ introduced in (2.4). Specifically, we set

$$[\xi]_x^\varepsilon(\mathbf{x}) = \langle \xi, \theta^\varepsilon(\mathbf{x} - \cdot) \rangle \quad \text{for any } \mathbf{x} \in \Omega,\ \mathrm{dist}[\mathbf{x}, \partial\Omega] > \varepsilon,$$

where the kernels θ^ε are the same as in (2.3).

For a positive integer k and $p \geq 1$, the *Sobolev space* $W^{k,p}(\Omega; R^N)$ consists of all (vector) functions $\mathbf{v} : \Omega \to R^N$ such that the distributional derivatives of each component v_i, $i = 1, \ldots, N$ of $\mathbf{v}$ up to order k belong to $L^p(\Omega)$. The norm on $W^{k,p}(\Omega, R^N)$ is given by formula

$$\|\mathbf{v}\|_{W^{k,p}(\Omega; R^N)} = \left(\sum_{|\mathbf{j}| \leq k} \sum_{i=1}^{N} \int_{\Omega} |\partial_x^{\mathbf{j}} u_i|^p\, \mathrm{d}\mathbf{x} \right)^{1/p} \tag{2.8}$$

with the obvious modification for $p = \infty$. Similarly, as above, we shall write $W^{k,p}(\Omega)$ for $W^{k,p}(\Omega; R)$.

Intuitively, the Sobolev spaces represent a completion of the set of smooth functions with respect to the L^p-norm of the derivatives up to order k. Indeed the following theorem holds.

Theorem 2.4 [1, Theorem 3.18] *Let a domain $\Omega \subset R^N$ possess a segment property, that is, for each $\mathbf{x} \in \partial\Omega$ there exists $r > 0$ and a vector $\mathbf{d}_x \in R^N$ such that if $\mathbf{y} \in \overline{\Omega}$, $|\mathbf{y} - \mathbf{x}| \leq r$, then $\mathbf{y} + s\mathbf{d}_x \in \Omega$ for all $s \in (0, 1)$.*

Then the set

$$C^\infty(\overline{\Omega}) \cap \{v \mid \|v\|_{W^{k,p}(\Omega)} < \infty\}$$

is dense in $W^{k,p}(\Omega)$, $p \in [1, \infty)$.

The spaces $W_0^{k,p}(\Omega; R^N)$, $1 \leq p < \infty$ are defined as the completion of $\mathcal{D}(\Omega; R^N)$ with respect to the norm $\|\cdot\|_{W^{k,p}(\Omega; R^N)}$. Unlike the larger spaces $W^{k,p}$, the space $W_0^{k,p}(\Omega; R^N)$ can always be considered as a subspace of $W^{k,p}(R^N, R^N)$ extending the functions to be zero outside Ω. Consequently, the shape of the boundary $\partial\Omega$ has no influence on the regularity properties of functions in $W_0^{k,p}$.

The dual spaces $[W_0^{k,p}(\Omega)]^*$ to $W_0^{k,p}(\Omega)$ are identified through the following theorem.

Theorem 2.5 [1, Theorem 3.10] *Let* $\Omega \subset R^N$ *be a domain, and let* $1 \leq p < \infty$.

Then any linear form $\xi \in [W_0^{k,p}(\Omega)]^*$ *admits a representation*

$$\langle \xi, v \rangle = \sum_{|\mathbf{j}| \leq k} \int_\Omega (-1)^{|\mathbf{j}|} w_{\mathbf{j}} \, \partial_{\mathbf{x}}^{\mathbf{j}} v \, d\mathbf{x} \quad \text{where } w_{\mathbf{j}} \in L^{p'}(\Omega), \quad \frac{1}{p} + \frac{1}{p'} = 1. \tag{2.9}$$

Moreover, the norm on the dual space can be given as

$$\|\xi\|_{[W^{k,p}(\Omega)]^*} = \inf \left\{ \max_{|\mathbf{j}| \leq k} \|w_{\mathbf{j}}\|_{L^{p'}(\Omega)} \Big| w_{\mathbf{j}}, |\mathbf{j}| \leq k \text{ satisfy } (2.9) \right\}.$$

Since $\mathcal{D}(\Omega)$ is dense (by definition) in $W_0^{k,p}(\Omega)$, $1 \leq p < \infty$, the dual $[W_0^{k,p}(\Omega)]^*$ can be understood as a subspace of the space of distributions $\mathcal{D}'(\Omega)$, which we denote as $W^{-k,p'}(\Omega)$.

One of most important properties of the Sobolev spaces are *imbedding theorems*. A fundamental result in this direction is the *Rellich–Kondrachov compactness theorem*

Theorem 2.6 [127, Theorem 2.5.1] *Let* $\Omega \subset R^N$ *be a bounded domain.*

(1) *Then, if* $kp < N$ *and* $p \geq 1$, *the space* $W_0^{k,p}(\Omega)$ *is continuously imbedded in* $L^q(\Omega)$ *for any*

$$1 \leq q \leq p^* = \frac{Np}{N - kp}.$$

Moreover, the imbedding is compact if $k > 0$ *and* $q < p^*$.

(2) *If* $kp = N$, *the space* $W_0^{k,p}(\Omega)$ *is compactly imbedded in* $L^q(\Omega)$ *for any finite* q.

(3) *If* $kp > N + \nu$, $\nu \geq 0$, *then* $W_0^{k,p}(\Omega)$ *is compactly imbedded in* $C^{0,\nu}(\overline{\Omega})$.

Here, one should remember that *a priori* the Sobolev spaces like the Lebesgue spaces consist of classes of equivalent functions. On the other hand, the functions in $C(\overline{\Omega})$ possess well-defined pointwise values. This rather ambiguous situation can be resolved, introducing the *precise representative* of a locally integrable function v as

$$v(\mathbf{x}) = \begin{cases} \lim_{h \to 0+} \frac{1}{|B_h(\mathbf{x})|} \int_{B_h(\mathbf{x})} v(\mathbf{y}) \, d\mathbf{y} & \text{if the limit exists,} \\ 0 & \text{otherwise,} \end{cases}$$

where $B_h(\mathbf{x})$ denotes the ball of radius $h > 0$ centered at $\mathbf{x} \in O$. A function $v \in L^1_{\text{loc}}(\Omega)$ coincides with its precise representative at any *Lebesgue point* $\mathbf{x} \in \Omega$. The set of all Lebesgue points of any locally integrable function is of full measure, that is, its complement in Ω is of measure zero. In fact, for functions belonging to a Sobolev space $W^{k,p}$, the "singular set" where the precise representative of a function v need not coincide with v is much "smaller" being of (k, p)-capacity zero (see e.g. [127]).

The conclusion of Theorem 2.6 is stated in terms of functions in $W_0^{k,p}$. A natural question is to identify the classes of domains for which the same result is valid for functions in $W^{k,p}(\Omega)$. It turns out that the answer is positive for domains having the *extension property* (more precisely (k, p)-extension property), specifically, there exists a bounded linear operator

$$\mathcal{L} : W^{k,p}(\Omega) \to W^{k,p}(R^N), \quad \mathcal{L}[v]|_\Omega = v.$$

A fundamental result of Calderon and Stein states that any Lipschitz domain has the extension property for any k, p. A domain $\Omega \subset R^N$ is called *Lipschitz* if its boundary can be locally described as a graph of a Lipschitz mapping defined on some open ball in R^{N-1}.

Theorem 2.7 [127, Remark 2.5.2] *The conclusion of Theorem 2.6 remains valid for the spaces $W^{k,p}(\Omega)$ provided the domain Ω has the (k, p)-extension property.*

As for the dual spaces, the following assertion results directly from Theorem 2.6.

Theorem 2.8 *Let $\Omega \subset R^N$ be a bounded domain. Let $k > 0$, and let $q \in R$ satisfy*

$$q > \frac{p^*}{p^* - 1}, \quad \text{where } p^* = \frac{Np}{N - kp} \quad \text{if } kp < N,$$

$$q > 1 \quad \text{if } kp = N,$$

or

$$q \geq 1 \quad \text{for } kp > N.$$

Then $L^q(\Omega)$ is compactly imbedded into the dual space $W^{-k,p'}(\Omega)$.

On the point of conclusion, we report a characterization of Sobolev functions in terms of integrability of their spatial shifts.

Theorem 2.9   [127, Theorem 2.1.6]   *Let $\Omega \subset R^N$ be a domain, and $K \subset \Omega$ compact.*
    Then

$$\int_K |v(\mathbf{x}+\mathbf{h}) - v(\mathbf{x})|^p \, d\mathbf{x} \leq c(p,\Omega)|h|^p \|v\|_{W^{1,p}(\Omega)}^p$$

for any $h \in R^N$ small enough, $v \in W^{1,p}(\Omega)$, $1 \leq p < \infty$.

2.1.4 Remarks

The functions we shall deal with represent physical quantities depending both on the time $t \in I$ and the spatial variable $\mathbf{x} \in \Omega$. Accordingly, a function $v = v(t,\mathbf{x})$ may be viewed either as a mapping $v : t \in I \mapsto v(t,\cdot)$ ranging in a suitable function space or as a (real) function of t and $\mathbf{x}$, as the case may be. We shall make no disctinction between these two descriptions.

2.2 Weak convergence

To establish the existence of a solution to a nonlinear system of partial differential equations, an obvious idea is to collect all available *a priori* estimates, that hold for any smooth solution of the problem; and to construct a sequence of "easier" approximate problems which are solvable and whose solutions will satisfy the same estimates. The crucial question to be answered then is whether or not the sequence of approximate solutions will converge to a solution of the original problem.

As we will see, the *a priori estimates* for the Navier–Stokes system result mostly from the total energy and entropy balance equations and are relatively poor in the sense that they ensure only boundedness of the density and other state variables in some Lebesgue space L^p. Accordingly, the main stumbling blocks in establishing rigorous existence results are the nonlinear constitutive equations for the pressure p, the specific heat energy Q, and the heat conductivity κ.

Many of the most important techniques set forth in recent years to deal with nonlinear problems are based on weak convergence methods. In order to characterize the incompatibility of weak convergence with nonlinear compositions, several tools have been developed as the slicing measures, the defect measures, the Young measures etc. (see e.g. Chapter 1 of [41]). Similar concepts will be used later in this book as well.

Let X be a Banach space and X^* its dual—the space of all bounded linear forms on X. The duality pairing between X^* and X will be denoted by $\langle,\rangle$.

The *weak-(*) topology* on X^* is generated by a system of semi-norms

$$p_v^*(f) = |\langle f, v \rangle|, \quad v \in X.$$

Similarly, one can define the *weak topology* on the space X through the system

$$p_f(v) = |\langle f, v \rangle|, \quad f \in X^*.$$

The space X^* endowed with the weak-(*) topology will be denoted as X^*_{weak} and, similarly, we shall write X_{weak} to stress the topological structure imposed on X by the weak topology.

The fundamental theorem of Banach–Alaoglu–Bourbaki states that any bounded ball in X^* is $\sigma(X^*, X)$-compact (see e.g. Theorem 7 of Chapter III in [68]). If, moreover, X is separable, then the weak-(*) topology is metrizable on bounded sets in X^*, and we can introduce the metric space $C(\overline{O}; X^*_{\text{weak}})$ of functions $v: \overline{O} \mapsto X^*$ which are continuous with respect to the weak topology. We have

$$v_n \to v \quad \text{in } C(\overline{O}; X^*_{\text{weak}}) \text{ if } \langle v_n(\mathbf{y}), \chi \rangle \to \langle v(\mathbf{y}), \chi \rangle$$

uniformly with respect to $\mathbf{y} \in \overline{O}$ for any $\chi \in X$.

The following assertion is a straightforward consequence of Theorem 2.1.

Corollary 2.1   *Let $\overline{O} \subset R^M$ be compact and let X be a separable Banach space. Assume that $v_n: \overline{O} \to X^*$, $n = 1, 2, \ldots$ is a sequence of measurable functions such that*

$$\operatorname*{ess\,sup}_{\mathbf{y} \in \overline{O}} \|v_n(\mathbf{y})\|_{X^*} \leq M \quad \text{uniformly in } n = 1, 2, \ldots$$

Moreover, let the family of (real) functions

$$\langle v_n, \Phi \rangle : \mathbf{y} \mapsto \langle v_n(\mathbf{y}), \Phi \rangle, \quad \mathbf{y} \in \overline{O}, \ n = 1, 2, \ldots$$

be equi-continuous for any fixed Φ belonging to a dense subset in the space X. Then $v_n \in C(\overline{O}; X^*_{\text{weak}})$ for any $n = 1, 2, \ldots$, and there exists $v \in C(\overline{O}; X^*_{\text{weak}})$ such that

$$v_n \to v \quad \text{in } C(\overline{O}, X^*_{\text{weak}}) \text{ as } n \to \infty$$

passing to a subsequence as the case may be.

2.2.1   Weak convergence of integrable functions

Now, we focus on functions belonging to the Lebesgue space $L^1(O)$. Consider a sequence $\{v_n\}_{n=1}^{\infty}$ which is bounded in $L^1(O)$, where $O \subset R^M$ is a bounded open set. It is well known that the space $L^1(O)$ is neither reflexive nor isomorphic to any dual so one cannot expect, in general, $\{v_n\}_{n=1}^{\infty}$ to converge weakly to some function $v \in L^1(O)$. However, any function v_n defines a bounded linear form on the space $C(\overline{O})$,

$$\langle v_n, \varphi \rangle = \int_O v_n \, \varphi \, \mathrm{d}\mathbf{y} \quad \text{for any } \varphi \in C(\overline{O});$$

that means that v_n can be understood as a *signed measure* on $\overline{O}$ (see Theorem 1 of Chapter 6.3 in [68]). Accordingly, $\{v_n\}_{n=1}^{\infty}$ contains a subsequence such that

$$v_n \to v \quad \text{weakly-(*)} \quad \text{in } \mathcal{M}(\overline{O})$$

where v is a uniquely determined distribution from the space $\mathcal{M}(\overline{O})$ of all signed measures on $\overline{O}$—the dual space to $C(\overline{O})$. As already pointed out, the limit measure v need not be absolutely continuous with respect to the Lebesgue measure, that is, v need not be representable by an integrable function.

A useful criterion when the limit v *does belong* to $L^1(O)$ is given in the following assertion.

Theorem 2.10   [40, Theorem 1.3, Chapter 8]   *Let* $\mathcal{V} \subset L^1(O)$, *where* $O \subset R^M$ *is a bounded measurable set.*

Then the following statements are equivalent:

(1) *any sequence* $\{v_n\}_{n=1}^{\infty} \subset \mathcal{V}$ *contains a subsequence weakly converging in* $L^1(O)$;

(2) *for any* $\varepsilon > 0$ *there exists* $k > 0$ *such that*

$$\int_{\{|v| \geq k\}} |v(\mathbf{y})| \, d\mathbf{y} \leq \varepsilon \quad \text{for all } v \in \mathcal{V};$$

(3) *for any* $\varepsilon > 0$ *there exists* $\delta > 0$ *such for all* $v \in \mathcal{V}$

$$\int_M |v(\mathbf{y})| \, d\mathbf{y} < \varepsilon$$

for any measurable set $M \subset O$ *such that*

$$|M| < \delta;$$

(4) *there exists a non-negative convex function* Φ,

$$\Phi : [0, \infty) \to [0, \infty), \quad \lim_{z \to \infty} \frac{\Phi(z)}{z} = \infty,$$

such that

$$\sup_{v \in \mathcal{V}} \int_O \Phi(|v(\mathbf{y})|) \, d\mathbf{y} \leq c.$$

Weak convergence in L^p*-spaces is nothing other than convergence of integral averages* A sequence $\{v_n\}_{n=1}^{\infty}$ converges weakly to v in $L^p(O)$, $1 \leq p < \infty$ if and only if

(1) $\|v_n\|_{L^p(O)}$ is bounded;

(2) $\int_M v_n(\mathbf{x}) \, d\mathbf{x} \to \int_M v(\mathbf{x}) \, d\mathbf{x}$ for any open set $M \subset O$.

In general, the operation of taking the integral average does not commute with a nonlinear composition represented by a superposition operator associated to a function g unless g is linear. If

$$v_n \to v \text{ weakly in } L^1(O), \quad O \subset R^M \text{ bounded,}$$

and $g : R \to R$ is a bounded function, then one can find a subsequence (not relabeled) such that

$$g(v_n) \to \overline{g(v)} \text{ weakly-(*) in } L^\infty(O), \quad \text{in particular, weakly in } L^1(O),$$

where, in general, $\overline{g(v)} \neq g(v)$. Here and in what follows, any weak L^1-limit of a composition $g(v_n)$, where $v_n \to v$ weakly in $L^1(O)$, will be denoted by $\overline{g(v)}$.

It might seem that for each g the choice of the subsequence $\{v_n\}_{n=1}^\infty$ for which $\{g(v_n)\}_{n=1}^\infty$ is weakly convergent depends on g. However, the following assertion holds.

Proposition 2.1   *Let $O \subset R^M$ be a bounded open set. Let $\{\mathbf{v}_n\}_{n=1}^\infty$ be a sequence of measurable functions,*

$$\mathbf{v}_n : O \to R^N,$$

such that

$$\sup_{n \geq 1} \int_O \Phi(|\mathbf{v}_n|) \, \mathrm{d}\mathbf{y} < \infty$$

for a certain continuous function $\Phi : [0, \infty) \to [0, \infty)$.
    Then there exists a subsequence (not relabeled) such that

$$g(\mathbf{v}_n) \to \overline{g(\mathbf{v})} \text{ weakly in } L^1(O)$$

for all continuous functions $g : R^N \to R$ satisfying

$$\lim_{|z| \to \infty} \frac{|g(z)|}{\Phi(|z|)} = 0. \tag{2.10}$$

Proof   (i) To begin with, observe that $\{g(\mathbf{v}_n)\}_{n=1}^\infty$ contains a subsequence weakly converging in $L^1(O)$. In view of Theorem 2.10, it is enough to show that the integrals

$$\int_{\{|g(\mathbf{v}_n)| \geq M\}} |g(\mathbf{v}_n)| \, \mathrm{d}\mathbf{y} \to 0 \quad \text{for } M \to \infty$$

uniformly in n. As g is continuous, we have

$$\{|g(\mathbf{v}_n)| \geq M\} \subset \{|\mathbf{v}_n| \geq k(M)\}, \quad \text{with } k(M) \to \infty \text{ for } M \to \infty.$$

Consequently,

$$\int_{\{g(\mathbf{v}_n) \geq M\}} |g(\mathbf{v}_n)| \, \mathrm{d}\mathbf{y} \leq \int_{\{|\mathbf{v}_n| \geq k(M)\}} |g(\mathbf{v}_n)| \, \mathrm{d}\mathbf{y}$$

$$\leq \sup_{|\mathbf{z}| \geq k(M)} \frac{|g(\mathbf{z})|}{\Phi(|\mathbf{z}|)} \int_O \Phi(|\mathbf{v}_n|) \, \mathrm{d}\mathbf{y},$$

where, in accordance with hypothesis (2.10), the right-hand side is small uniformly for large M.

(ii) The family $\{p\}_{p\in\mathcal{P}}$ of all polynomials on R^N with rational coefficients is dense in the space $C(B)$ for any bounded ball $B \subset R^N$.

Now, we introduce the *cut-off functions* T_k,

$$T_k(z) \equiv kT\left(\frac{z}{k}\right), \tag{2.11}$$

where $T \in C^\infty(R)$ is chosen so that

$$T(z) \equiv \begin{cases} T(z) = z & \text{for } z \in [0,1], \\ T(z) \text{ concave} & \text{on } [0,\infty), \\ T(z) = 2 & \text{for } z \geq 3, \\ T(z) = -T(-z) & \text{for } z \in (-\infty, 0]. \end{cases} \tag{2.12}$$

Similarly, we set

$$\chi_m(\mathbf{z}) = \chi\left(\frac{\mathbf{z}}{m}\right),$$

where

$$\chi \in \mathcal{D}(R^N), \quad 0 \leq \chi \leq 1, \quad \chi(\mathbf{z}) = 1 \quad \text{for } |\mathbf{z}| \leq 1, \quad \chi(\mathbf{z}) = 0 \quad \text{if } |\mathbf{z}| \geq 2.$$

Using a simple diagonalization procedure we can find a subsequence such that

$$\chi_m(\mathbf{v}_n)T_k(p(\mathbf{v}_n)) \;\rightarrow\; \overline{\chi_m(\mathbf{v})T_k(p(\mathbf{v}))} \quad \text{weakly-(*) in } L^\infty(O) \tag{2.13}$$

for all $p \in \mathcal{P}$, $k = 1, 2, \ldots$, and $m = 1, 2, \ldots$.

Now, let g be a continuous function satisfying (2.10). We can write

$$\int_O g(\mathbf{v}_n)\varphi\,\mathrm{d}\mathbf{y} = \int_O \chi_m(\mathbf{v}_n)g(\mathbf{v}_n)\varphi\,\mathrm{d}\mathbf{y} + \int_O (1-\chi_m(\mathbf{v}_n))g(\mathbf{v}_n)\varphi\,\mathrm{d}\mathbf{y},$$

where

$$\left|\int_O (1-\chi_m(\mathbf{v}_n))g(\mathbf{v}_n)\varphi\,\mathrm{d}\mathbf{y}\right| \leq \int_{\{|\mathbf{v}_n|\geq m\}} |g(\mathbf{v}_n)||\varphi|\,\mathrm{d}\mathbf{y}$$

$$\leq \|\varphi\|_{L^\infty(O)} \sup_{|\mathbf{z}|\geq m} \frac{|g(\mathbf{z})|}{\Phi(|\mathbf{z}|)} \int_O \Phi(|\mathbf{v}_n|)\,\mathrm{d}\mathbf{y}.$$

In accordance with hypothesis (2.10), the most right expression in the above formula is uniformly small provided m has been chosen large enough.

On the other hand, given $m > 0$ and $\varepsilon > 0$, one can find $p \in \mathcal{P}$ and $k \geq 1$ such that

$$\sup_{-2m \leq |\mathbf{z}| \leq 2m} |T_k(p(\mathbf{z})) - g(\mathbf{z})| \leq \varepsilon.$$

Consequently,

$$\int_O \chi_m(\mathbf{v}_n) g(\mathbf{v}_n) \varphi \, d\mathbf{y} = \int_O \chi_m(\mathbf{v}_n) T_k(p(\mathbf{v}_n)) \varphi \, d\mathbf{y}$$
$$+ \int_O \chi_m(\mathbf{v}_n)(g(\mathbf{v}_n) - T_k(p(\mathbf{v}_n))) \varphi \, d\mathbf{y},$$

where

$$\left| \int_O \chi_m(\mathbf{v}_n)(g(\mathbf{v}_n) - T_k(p(\mathbf{v}_n))) \varphi \, d\mathbf{y} \right| \leq \varepsilon \|\varphi\|_{L^\infty(O)} |O|.$$

As ε can be chosen arbitrarily small, the above relations imply convergence of the sequence

$$\int_O g(\mathbf{v}_n) \varphi \, d\mathbf{y}, n = 1, 2, \dots$$

for any test function $\varphi \in L^\infty(O)$.

$\square$

2.2.2 Convex functions

An important class of composition operators we shall deal with is represented by *convex functions*.

More specifically, we consider a function $\Phi : R^M \mapsto (-\infty, \infty]$ which is *lower semi-continuous* and *convex*, that is, the set

$$\text{epi}[\Phi] \equiv \{ (\mathbf{z}, a) \in R^M \times R \mid a \geq \Phi(\mathbf{z}) \}$$

is closed and convex (cf. Theorems 2.1, 2.3 of Chapter 1 in [40]). When dealing with convex functions, it is convenient to admit the value $\Phi(\mathbf{z}) = \infty$ for certain $\mathbf{z} \in R^M$.

Moreover, we shall say that Φ is *strictly convex* on a convex set $U \subset R^M$ if

$$\Phi(\lambda \mathbf{w} + (1 - \lambda)\mathbf{z}) < \lambda \Phi(\mathbf{w}) + (1 - \lambda)\Phi(\mathbf{z})$$

for any $\mathbf{w}, \mathbf{z} \in U$, $\mathbf{w} \neq \mathbf{z}$, and any $\lambda \in (0, 1)$.

Convex functions have a remarkable property of being lower semi-continuous with respect to the weak topology on $L^1(O)$, as shown in the following assertion.

Theorem 2.11 *Let $O \subset R^N$ be a measurable set and $\{\mathbf{v}_n\}_{n=1}^{\infty}$ a sequence of functions in $L^1(O; R^M)$ such that*

$$\mathbf{v}_n \to \mathbf{v} \text{ weakly in } L^1(O; R^M).$$

Let $\Phi : R^M \to (-\infty, \infty]$ be a lower semi-continuous convex function such that $\Phi(\mathbf{v}_n) \in L^1(O)$ for any n, and

$$\Phi(\mathbf{v}_n) \to \overline{\Phi(\mathbf{v})} \text{ weakly in } L^1(O).$$

Then

$$\Phi(\mathbf{v}) \leq \overline{\Phi(\mathbf{v})} \quad a.a. \quad on \ O. \tag{2.14}$$

If, moreover, Φ is strictly convex on an open convex set $U \subset R^M$, and

$$\Phi(\mathbf{v}) = \overline{\Phi(\mathbf{v})} \quad a.a. \quad on \ O,$$

then

$$\mathbf{v}_n(\mathbf{y}) \to \mathbf{v}(\mathbf{y}) \quad \text{for a.a. } \mathbf{y} \in \{\mathbf{y} \in O \mid \mathbf{v}(\mathbf{y}) \in U\} \tag{2.15}$$

extracting a subsequence as the case may be.

Proof (i) Any convex lower semi-continuous function with values in $(-\infty, \infty]$ can be written as a supremum of its affine minorants:

$$\Phi(\mathbf{z}) = \sup\{a(\mathbf{z}) \mid a \text{ an affine function on } R^M, a \leq \Phi \text{ on } R^M\} \tag{2.16}$$

(see Theorem 3.1 of Chapter 1 in [40]). Recall that a function is called *affine* if it can be written as a sum of a linear and a constant function.

On the other hand, if $B \subset O$ is a measurable set, we have

$$\int_B \overline{\Phi(\mathbf{v})} \, d\mathbf{y} = \lim_{n \to \infty} \int_B \Phi(\mathbf{v}_n) \, d\mathbf{y} \geq \lim_{n \to \infty} \int_B a(\mathbf{v}_n) \, d\mathbf{y} = \int_B a(\mathbf{v}) \, d\mathbf{y}$$

for any affine function $a \leq \Phi$. Consequently,

$$\overline{\Phi(\mathbf{v})}(\mathbf{y}) \geq a(\mathbf{v})(\mathbf{y})$$

for any $\mathbf{y} \in O$ that is a Lebesgue point of both $\overline{\Phi(\mathbf{v})}$ and $\mathbf{v}$.

Thus formula (2.16) yields (2.14).

(ii) As any open set $U \subset R^M$ can be expressed as a countable union of compacts, it is enough to show (2.15) for

$$\mathbf{y} \in M_K \equiv \{\mathbf{y} \in O \mid \mathbf{v}(\mathbf{y}) \in K\},$$

where $K \subset U$ is compact.

Since Φ is strictly convex on U, there exists an open set V such that

$$K \subset V \subset \overline{V} \subset U,$$

and $\Phi : \overline{V} \to R$ is a Lipschitz function (see Corollary 2.4 of Chapter I in [40]). In particular, the subdifferrential $\partial\Phi(\mathbf{v})$ is non-empty for each $\mathbf{v} \in K$, and we have

$$\Phi(\mathbf{w}) - \Phi(\mathbf{v}) \geq \underline{\partial}\Phi(\mathbf{v}) \cdot (\mathbf{w} - \mathbf{v}) \quad \text{for any } \mathbf{w} \in R^M, \ \mathbf{v} \in K,$$

where $\underline{\partial}\Phi(\mathbf{v})$ denotes the linear form in the subdifferential $\partial\Phi(\mathbf{v}) \subset (R^M)^*$ with the smallest norm (see Corollary 2.4 of Chapter 1 in [40]).

Next, we shall show the existence of a function ω,

$$\omega \in C[0, \infty), \ \omega(0) = 0,$$

$$\tag{2.17}$$

ω non-decreasing on $[0, \infty)$ and strictly positive on $(0, \infty)$,

such that

$$\Phi(\mathbf{w}) - \Phi(\mathbf{v}) \geq \underline{\partial}\Phi(\mathbf{v}) \cdot (\mathbf{w} - \mathbf{v}) + \omega(|\mathbf{w} - \mathbf{v}|) \quad \text{for all } \mathbf{w} \in \overline{V}, \mathbf{v} \in K. \tag{2.18}$$

Were (2.18) not true, we would be able to find two sequences $\mathbf{w}_n \in \overline{V}$, $\mathbf{z}_n \in K$ such that

$$\Phi(\mathbf{w}_n) - \Phi(\mathbf{z}_n) - \underline{\partial}\Phi(\mathbf{z}_n) \cdot (\mathbf{w}_n - \mathbf{z}_n) \to 0 \quad \text{for } n \to \infty$$

while

$$|\mathbf{w}_n - \mathbf{z}_n| \geq \delta > 0 \quad \text{for all } n = 1, 2, \ldots.$$

Moreover, as K is compact, one can assume

$$\mathbf{z}_n \to \mathbf{z} \in K, \ \Phi(\mathbf{z}_n) \to \Phi(\mathbf{z}), \ \mathbf{w}_n \to \mathbf{w} \text{ in } \overline{V}, \ \underline{\partial}\Phi(\mathbf{z}_n) \to L \in R^M,$$

and, consequently,

$$\Phi(\mathbf{y}) - \Phi(\mathbf{z}) \geq L \cdot (\mathbf{y} - \mathbf{z}) \quad \text{for all } \mathbf{y} \in R^M,$$

that is, $L \in \partial\Phi(\mathbf{z})$.

Now, the function

$$\Psi(\mathbf{y}) \equiv \Phi(\mathbf{y}) - \Phi(\mathbf{z}) - L \cdot (\mathbf{y} - \mathbf{z})$$

is non-negative, convex, and

$$\Psi(\mathbf{z}) = \Psi(\mathbf{w}) = 0, \quad |\mathbf{w} - \mathbf{z}| \geq \delta.$$

Consequently, Ψ vanishes on the whole segment $[\mathbf{z}, \mathbf{w}]$, which is impossible as Φ is strictly convex on U.

Seeing that the function

$$a \mapsto \Phi(\mathbf{z} + a\mathbf{y}) - \Phi(\mathbf{z}) - a\underline{\partial}\Phi(\mathbf{z}) \cdot \mathbf{y}$$

is non-negative, convex, and non-decreasing for $a \in [0, \infty)$, we infer that the estimate (2.18) holds without the restriction $\mathbf{w} \in \overline{V}$. More precisely, there exists

ω as in (2.17) such that

$$\Phi(\mathbf{w}) - \Phi(\mathbf{v}) \geq \underline{\partial}\Phi(\mathbf{v}) \cdot (\mathbf{w} - \mathbf{v}) + \omega(|\mathbf{w} - \mathbf{v}|) \quad \text{for all } \mathbf{w} \in R^M, \ \mathbf{v} \in K. \quad (2.19)$$

Taking $\mathbf{w} = \mathbf{v}_n(\mathbf{y})$, $\mathbf{v} = \mathbf{v}(\mathbf{y})$ in (2.19) and integrating over the set M_K we get

$$\int_{M_K} \omega(|\mathbf{v}_n - \mathbf{v}|) \, d\mathbf{y} \leq \int_{M_K} \Phi(\mathbf{v}_n) - \Phi(\mathbf{v}) - \underline{\partial}\Phi(\mathbf{v}) \cdot (\mathbf{v}_n - \mathbf{v}) \, d\mathbf{y},$$

where the right-hand side tends to zero for $n \to \infty$. Note that the function $\underline{\partial}\Phi(\mathbf{v})$ is bounded measurable on M_k as Φ is Lipschitz on $\overline{V}$, and

$$\underline{\partial}\Phi(\mathbf{v}) = \lim_{\varepsilon \to 0} \nabla\Phi_\varepsilon(\mathbf{v}) \quad \text{for any } \mathbf{v} \in V,$$

where

$$\Phi_\varepsilon(\mathbf{v}) \equiv \min_{\mathbf{z} \in R^M} \left\{ \frac{1}{\varepsilon} |\mathbf{z} - \mathbf{v}| + \Phi(\mathbf{z}) \right\} \quad (2.20)$$

is a convex, continuously differentiable function on R^M (see Propositions 2.6, 2.11 of Chapter 2 in [14]).

Thus

$$\int_{M_K} \omega(|\mathbf{v}_n - \mathbf{v}|) \, d\mathbf{y} \to 0 \quad \text{for } n \to \infty$$

which yields pointwise convergence (for a subsequence) of $\{\mathbf{v}_n\}_{n=1}^\infty$ to $\mathbf{v}$ a.a. on M_K.

$\square$

Corollary 2.2　　*Let $O \subset R^M$ be a bounded measurable set, $\{\mathbf{v}_n\}_{n=1}^\infty$ a sequence of functions such that*

$$\mathbf{v}_n \to \mathbf{v} \text{ weakly in } L^1(O; R^N).$$

Let $\Phi : R^N \to (-\infty, \infty]$ be a convex lower semi-continuous function.
　　Then $\Phi(\mathbf{v}) : O \to R$ is integrable, and

$$\int_O \Phi(\mathbf{v}) \, d\mathbf{y} \leq \liminf_{n \to \infty} \int_O \Phi(\mathbf{v}_n) \, d\mathbf{y}$$

Proof　As already observed, Φ admits an affine minorant on R^N. Consequently, as O is bounded, we may assume $\Phi \geq 0$ on R^N. Moreover, it is enough to consider functions that are not identically equal to ∞.

Let us take the regularized functions Φ_ε introduced in (2.20). We have

$$0 \leq \Phi_\varepsilon(\mathbf{v}) \leq c(\varepsilon)(1 + |\mathbf{v}|) \quad \text{for all } \mathbf{v} \in R^N;$$

whence, by virtue of Theorem 2.10,

$$\Phi_\varepsilon(\mathbf{v}_n) \to \overline{\Phi_\varepsilon(\mathbf{v})} \text{ weakly in } L^1(O) \text{ for } n \to \infty$$

at least for a subsequence as the case may be.

Consequently, utilizing Theorem 2.11 we have

$$\int_O \Phi_\varepsilon(\mathbf{v})\,\mathrm{d}y \le \int_O \overline{\Phi_\varepsilon(\mathbf{v})}\,\mathrm{d}y = \lim_{n\to\infty}\int_O \Phi_\varepsilon(\mathbf{v}_n)\,\mathrm{d}y \le \liminf_{n\to\infty}\int_O \Phi(\mathbf{v}_n)\,\mathrm{d}y.$$

In order to conclude, it is enough to observe that

$$\Phi_\varepsilon(\mathbf{v}) \nearrow \Phi(\mathbf{v}) \quad \text{for all } \mathbf{v} \in R^N$$

(see Proposition 2.11 of Chapter 2 in [14]).

$\square$

2.3 Vector functions of one real variable

All physical quantities we shall deal with are functions of the time t and the position $\mathbf{x}$. In accordance with Chapter 1, the state of a thermodynamics system (fluid) at time t is fully characterized by the *instantaneous value* of the macroscopic variables $\varrho(t)$, $\mathbf{u}(t)$, and $\vartheta(t)$ or, equivalently, $\varrho(t)$, $(\varrho\mathbf{u})(t)$, and $\varrho Q(\vartheta)(t)$. On the other hand, these quantities will be numerical functions of t and $\mathbf{x}$ belonging to the Lebegue space L^1, in particular, their values are defined only for almost any t in the sense of the Lebesgue measure on a given time interval.

To fix ideas, we take the "initial time" $t_I = 0$ and consider a time interval $I = (0, T)$. All scalar physical quantities we shall deal with will be representable by a vector valued function v:

$$v : [0, T] \to L^1(\Omega),\ \Omega \subset R^N \text{ a bounded domain,}$$

$$v \text{ measurable on } [0, T];\ t \mapsto \|v(t)\|_{L^1(\Omega)} \text{ integrable on } (0, T);$$

in other words, $v \in L^1(0, T; L^1(\Omega))$.

In particular, since $v(t) \in L^1(\Omega)$ for a.a. $t \in [0, T]$ and $L^1(\Omega)$ is a separable space, v is measurable if and only if the (real valued) functions

$$t \mapsto \int_\Omega v(t)\eta\,\mathrm{d}x \text{ are measurable for any } \eta \in L^\infty(\Omega). \tag{2.21}$$

As a matter of fact, it is enough to verify (2.21) only for smooth functions $\eta \in \mathcal{D}(\Omega)$. Indeed for any $\eta \in L^\infty(\Omega)$ there exists a sequence $\eta_n \in \mathcal{D}(\Omega)$ such that

$$\|\eta_n\|_{L^\infty(\Omega)} \le \|\eta\|_{L^\infty(\Omega)} \quad \text{for all } n = 1, 2, \ldots,\ \eta_n \to \eta \text{ a.a. on } \Omega.$$

Thus, by virtue of the Lebesgue theorem,

$$\int_\Omega v(t)\eta_n\,\mathrm{d}x \to \int_\Omega v(t)\eta\,\mathrm{d}x \text{ for a.a. } t \in (0, T);$$

whence the function $t \mapsto \int_\Omega v(t)\eta\,\mathrm{d}x$, being a pointwise limit of measurable functions, is measurable together with $t \mapsto \int_\Omega v(t)\eta_n\,\mathrm{d}x$.

The *instantaneous values* of integrable functions are well defined at the *Lebesgue points*. Let $v \in L^1(0, T; X)$ be a function ranging in a Banach space X.

A point $\tau \in [0, T)$ is called a *right Lebesgue point* of v if there exists $m \in X$ such that

$$\lim_{h \to 0+} \frac{1}{h} \int_\tau^{\tau+h} \|v(t) - m\|_X \, \mathrm{d}t = 0.$$

Similarly, one can define a *left Lebesgue point* $\tau \in (0, T]$ through

$$\lim_{h \to 0+} \frac{1}{h} \int_{\tau-h}^{\tau} \|v(t) - m\|_X \, \mathrm{d}t = 0.$$

The set of all right (left) Lebesgue points of an integrable function is of full measure in $(0, T)$, and $m = v(\tau)$ for a.a. $\tau \in (0, T)$ (see e.g. Appendix in [14]). This motivates the following definition.

Definition 2.1   Let $v \in L^1(0, T; L^1(\Omega))$ be a vector valued function. The *right instantaneous value* of v at $\tau \in [0, T)$ is a measure $v(\tau+) \in \mathcal{M}(\overline{\Omega})$,

$$\langle v(\tau+), \eta \rangle \equiv \begin{cases} \lim_{h \to 0+} \frac{1}{h} \int_\tau^{\tau+h} \int_\Omega v(t, \cdot) \eta \, \mathrm{d}\mathbf{x} \, \mathrm{d}t \\ \quad \text{if the limit exists for any } \eta \in C(\overline{\Omega}), \qquad (2.22) \\ 0 \quad \text{otherwise.} \end{cases}$$

Similarly, for $\tau \in (0, T]$, the *left instantaneous value* $v(\tau-)$ is defined through

$$\langle v(\tau-), \eta \rangle \equiv \begin{cases} \lim_{h \to 0+} \frac{1}{h} \int_{\tau-h}^{\tau} \int_\Omega v(t, \cdot) \eta \, \mathrm{d}\mathbf{x} \, \mathrm{d}t \\ \quad \text{if the limit exists for any } \eta \in C(\overline{\Omega}), \qquad (2.23) \\ 0 \quad \text{otherwise.} \end{cases}$$

The *instantaneous value* $v(\tau)$ is defined as

$$v(\tau) \equiv \frac{v(\tau-) + v(\tau+)}{2} \quad \text{for any } \tau \in (0, T). \qquad (2.24)$$

Note that, by virtue of the Banach–Steinhaus theorem, the limits in (2.22), (2.23) define a bounded linear functional on $C(\overline{\Omega})$, that is, a signed measure on $\overline{\Omega}$, as soon as they exist for any $\eta \in C(\overline{\Omega})$. Moreover, for a.a. $\tau \in (0, T)$, this measure is absolutely continuous with respect to the Lebesgue measure and coincides with $v(\tau) \in L^1(\Omega)$.

To conclude this section, we recall a variant of the *Gronwall–Bellman theorem*.

Lemma 2.1   [74, Lemma 4.3.1]
    Let $h \in L^\infty(0, T)$, $h \geq 0$, $a \in R$, and $b \in L^1(0, T)$, $b \geq 0$ *satisfy the inequality*

$$h(\tau) \leq a + \int_0^\tau b(t) h(t) \, \mathrm{d}t \quad \text{for all } \tau \in [0, T].$$

Then

$$h(\tau) \le a \, \exp \left(\int_0^\tau b(t) \, \mathrm{d}t \right) \quad \textit{for a.a. } \tau \in [0, T].$$

2.4 Bibliographical notes

2.1 There is a vast literature available concerning the function spaces introduced in Section 2.1. In particular, the Lebesgue and Sobolev spaces were studied by Adams [1], Maz'ya [94], Triebel [114], among others. For applications to concrete partial differential equations see for instance [82], [97]. Pointwise properties of Sobolev functions were studied by Ziemer [127]. A more advanced approach including spaces defined via interpolation methods can be found in the monograph [113].

2.2 The weak convergence methods and their applications to partial differential equations are reviewed in the monograph [41] by Evans. A more classical material including the introduction of the Young measures can be found in the paper [111] by Tartar. Another version of the fundamental theorem for Young measures which is strongly related to Proposition 2.1 was proved by Ball in [6]. Similar results can be found in [24], [41]. DiPerna and Majda [33] use defect measures to describe possible concentrations in sequences of solutions to the Euler equations. A refined concept of the so-called H-meaures was introduced by Tartar [112] and Gerard [60].

The material concerning the convex functions is basically taken from the monograph of Ekeland and Temam [40]. Other useful facts may be also found in [14].

2.3 The classical reference material for the theory of vector valued functions ranging in an infinite dimensional space is represented by the monographs on functional analysis like those of Dunford and Schwartz [37], Yosida [125] etc. The theory is nicely summarized in the Appendix of [14]. More information is available in [57].

3

A PRIORI ESTIMATES

In a broad sense, *a priori estimates* are bounds imposed on any smooth solution of a given problem by the equations, the boundary and initial conditions, and other constraints as the case may be. *A priori* estimates play a crucial role in the analysis of any *nonlinear* system of equations because they determine the function spaces the solution is expected to live in. These must be at least strong enough to give meaning to all nonlinear terms in the equations, and, if possible, to ensure the so-called stability (or compactness) of the set of all solutions. This means that any sequence of solutions that is bounded in the function spaces determined by *a priori* estimates converges in a certain sense to another solution of the same problem. If a nonlinear problem enjoys this property, there is usually a good chance of having a rigorous existence theory.

The whole of this chapter is devoted to *a priori* estimates for the Navier–Stokes system of equations (1.34), (1.35), (1.36), where the state variables $\varrho, \mathbf{u}, \vartheta$ are interrelated through the constitutive equations (1.37)–(1.40) and satisfy the boundary conditions (1.41), (1.42).

A priori estimates hold for any *regular* (smooth) solution of the problem. Accordingly, we shall assume that the spatial domain $\Omega \subset R^N$ belongs to the class $C^{2+\nu}, \nu > 0$.

Next, the density ϱ will be assumed to be continuously differentiable with partial derivatives $\partial_t \varrho, \partial_{x_i} \varrho, i = 1, \ldots, N$ continuous and bounded on the domain $(0, T) \times \Omega$. Moreover, ϱ is strictly positive at the initial instant $t = 0$. More precisely, we assume there are constants $\underline{\varrho}, \overline{\varrho}$ such that

$$0 < \underline{\varrho} \leq \lim_{t \to 0+} \varrho(t, \mathbf{x}) \leq \overline{\varrho} \quad \text{for all } \mathbf{x} \in \Omega.$$

Similarly, we suppose that the fluid velocity $\mathbf{u}$ together with $\partial_t \mathbf{u}$ and all spatial derivatives up to order 2 are continuous and bounded on $(0, T) \times \Omega$; $\mathbf{u}$ vanishes on the boundary $\partial \Omega$:

$$\lim_{\mathbf{x} \to \partial \Omega} \mathbf{u}(t, \mathbf{x}) = 0 \quad \text{for any } t \in (0, T).$$

The temperature $\vartheta, \partial_t\vartheta$, and all spatial derivatives of ϑ up to order 2 will be continuous and bounded on $(0,T) \times \Omega$; ϑ is strictly positive for $t = 0$:

$$0 < \underline{\vartheta} \leq \lim_{t\to0+} \vartheta(t,\mathbf{x}) \quad \text{for all } \mathbf{x} \in \Omega;$$

ϑ satisfies the boundary conditions (1.42):

$$\lim_{\mathbf{y}\to\mathbf{x}} \nabla_x\vartheta(\mathbf{y}) \cdot \mathbf{n}(\mathbf{x}) = 0 \quad \text{for any } t \in (0,T) \text{ and } \mathbf{x} \in \partial\Omega.$$

Finally, the functions $\varrho, \mathbf{u}$, and ϑ will be assumed to solve the equation of continuity (1.34), the system of momentum equations (1.35), and the thermal energy equation (1.36) at any point $(t, \mathbf{x}) \in (0,T) \times \Omega$.

The nonlinear functions appearing in the constitutive equations for the *pressure*, the *thermal conductivity*, and the *thermal energy* will satisfy certain structural conditions specified as below:

(i) the pressure

$$p(\varrho, \vartheta) = p_e(\varrho) + \vartheta p_\vartheta(\varrho),$$

where

$$\left\{\begin{array}{l} p_e \in C[0,\infty) \cap C^1(0,\infty), \quad p_e(0) = 0, \\ p_e'(\varrho) \geq a_1\varrho^{\gamma-1} - b \quad \text{for all } \varrho > 0, \\ p_e(\varrho) \leq a_2\varrho^{\gamma} + b \quad \text{for all } \varrho \geq 0, \end{array}\right\} \tag{3.1}$$

for certain constants $\gamma > 1, a_1 > 0, a_2$, and b,

$$\left\{\begin{array}{c} p_\vartheta \in C[0,\infty) \cap C^1(0,\infty), \quad p_\vartheta(0) = 0, \\ p_\vartheta \text{ is a non-decreasing function of } \varrho \in [0,\infty), \\ p_\vartheta(\varrho) \leq c(1 + \varrho^\Gamma) \quad \text{for all } \varrho \geq 0, \end{array}\right\} \tag{3.2}$$

where

$$\Gamma < \frac{\gamma}{2} \text{ if } N = 2, \quad \Gamma = \frac{\gamma}{N} \text{ for } N \geq 3;$$

(ii) the heat conductivity coefficient

$$\kappa = \kappa(\vartheta) \text{ belongs to the class } C^2[0,\infty),$$

$$k_1(\vartheta^\alpha + 1) \leq \kappa(\vartheta) \leq k_2(\vartheta^\alpha + 1) \quad \text{for all } \vartheta \geq 0, \tag{3.3}$$

with constants

$$k_1 > 0 \quad \text{and} \quad \alpha \geq 2; \tag{3.4}$$

(iii) the thermal energy

$$Q = Q(\vartheta) = \int_0^\vartheta c_v(z)\,\mathrm{d}z,$$

where

$$c_v \in C^1[0, \infty), \quad \inf_{z \in [0, \infty)} c_v(z) > 0, \tag{3.5}$$

and

$$c_v(\vartheta) \le c(1 + \vartheta^{(\alpha/2)-1}). \tag{3.6}$$

Note that these hypotheses agree with the physical concepts discussed in Section 1.3.

Next, we shall assume that both the total energy and the entropy are bounded at time $t = 0$, specifically,

$$\limsup_{t \to 0+} E(t) \le E_0, \quad \liminf_{t \to 0+} S(t) \ge S_0, \tag{3.7}$$

where

$$E = E[\varrho, \mathbf{u}, \vartheta] \equiv \int_\Omega \frac{1}{2}\varrho|\mathbf{u}|^2 + \varrho P_e(\varrho) + \varrho Q(\vartheta)\mathrm{d}\mathbf{x}, \tag{3.8}$$

and

$$S = S[\varrho, \vartheta] \equiv \int_\Omega \varrho s(\varrho, \vartheta)\mathrm{d}\mathbf{x}, \quad \text{with } s(\varrho, \vartheta) = \int_1^\vartheta \frac{c_v(z)}{z}\,\mathrm{d}z - P_\vartheta(\varrho), \tag{3.9}$$

where the thermal pressure potential P_ϑ is defined by formula (1.25).

Finally, the amplitude of the external force will be denoted as B_f,

$$B_f \equiv \operatorname*{ess\,sup}_{t \in R,\ \mathbf{x} \in \Omega} |\mathbf{f}(t, \mathbf{x})|. \tag{3.10}$$

3.1 Estimates based on the maximum principle

There are estimates that can be seen directly from the fact that certain partial derivatives of the solution satisfy certain relations given by the system of equations in question. The best-known example is the so-called *maximum (or minimum) principle*, which holds for parabolic equations of "heat equation" type and is directly inherited by scalar conservation laws like the continuity equation (1.34).

As the velocity field is assumed to be regular, the system (1.1) introduced in Section 1.1 can be resolved to yield the *motion* $\mathbf{X}$.

In accordance with the continuity equation (1.34), the density ϱ computed along streamlines (characteristics) satisfies

$$\frac{d}{dt}\varrho(t, \mathbf{X}(t)) = -\varrho(t, \mathbf{X}(t))\operatorname{div}_x \mathbf{u}(t, \mathbf{X}(t)), \quad \mathbf{X}(0) = \mathbf{x}, \ t \in [0, T].$$

Now, a direct application of Gronwall's lemma (Lemma 2.1) yields

$$\underline{\varrho}\exp\left(-\int_0^\tau \|\operatorname{div}_x \mathbf{u}(t)\|_{L^\infty(\Omega)}\,dt\right) \le \varrho(\tau, \mathbf{x})$$

$$\le \overline{\varrho}\exp\left(\int_0^\tau \|\operatorname{div}_x \mathbf{u}(t)\|_{L^\infty(\Omega)}\,dt\right) \tag{3.11}$$

for any $\mathbf{x} \in \Omega$ and any $\tau \in [0, T]$.

Formula (3.11) would yield very strong pointwise estimates of the density in terms of initial data if we could show boundedness of $\operatorname{div}_x\mathbf{u}$ in the space $L^1(0, T; L^\infty(\Omega))$. Unfortunately, a uniform *a priori* bound on $\operatorname{div}_x\mathbf{u}$ is one of the major open problems at least for $N \ge 2$. We shall come back to this interesting issue when discussing the so-called pressure estimates in Chapter 5.

Thus the only useful information, which is independent of the amplitude of $\operatorname{div}_x\mathbf{u}$, is the fact that the density remains non-negative for any positive time provided it was the case for $t = 0$:

$$\varrho(t, \cdot) \ge 0 \quad \text{for any } t \ge 0 \text{ provided } \varrho(0, \cdot) \ge 0. \tag{3.12}$$

The lack of uniform estimates for ϱ from below results in serious mathematical difficulties when dealing with equations (1.35) and (1.36). Note that the leading terms $\partial_t(\varrho\mathbf{u})$ and $\partial_t(\varrho Q(\vartheta))$ "disappear" in the region where ϱ vanishes, the resulting system being rather "elliptic" than "parabolic"; in particular, it changes type.

We investigate next similar estimates for the temperature ϑ. In accordance with (1.41), (1.42), the velocity $\mathbf{u}$, together with the heat flux $\mathbf{q}$, vanishes on the boundary $\partial\Omega$. Thus one can apply the maximum principle to the parabolic equation (1.36) to deduce

$$\vartheta(t, \cdot) \ge 0 \quad \text{for any } t \ge 0 \text{ if } \vartheta(0, \cdot) \ge 0. \tag{3.13}$$

To make this argument more rigorous, it is enough to multiply (1.36) by an expression $G'(\vartheta)$, where $G : R \to R$ is a convex non-increasing function, and integrate by parts to obtain

$$\frac{d}{dt}\int_\Omega \varrho H(\vartheta)\,d\mathbf{x} \le \int_\Omega \left|p_\vartheta(\varrho)\operatorname{div}_x\mathbf{u}\right||G'(\theta)\theta|\,d\mathbf{x} \quad \text{where } H'(z) \equiv c_v(z)G'(z).$$

Indeed observe that, using the continuity equation (1.34), we can write

$$\partial_t(\varrho Q(\vartheta)) + \operatorname{div}_x(\varrho Q(\vartheta)\mathbf{u}) = \varrho(\partial_t Q(\vartheta) + \nabla_x Q(\vartheta) \cdot \mathbf{u}).$$

Under hypothesis (3.5), c_v is strictly positive on $[0, \infty)$, and one can choose $G(\vartheta) = \min\{-\vartheta, 0\}$ to deduce, utilizing Gronwall's lemma again, that

$G(\vartheta)(t) \equiv 0$ provided $G(\vartheta)$ vanishes initially, which implies (3.13). In fact, a more refined argument shows that ϑ rests strictly positive on $[0, T] \times \Omega$ as long as we deal with classical solutions.

Continuing our discussion concerning the physical interpretation of estimates (3.12), (3.13), we have to admit that there is no physical reason for the temperature ϑ to attain absolute zero $\vartheta = 0$. On the other hand, to our best knowledge, there are no suitable *a priori* estimates to prevent this phenomenon from occuring. However, as we shall see below, it can be shown that the set of these "singular points" is of Lebesgue measure zero.

3.2 Total mass conservation

In light of arguments discussed in Section 1.1, it is intuitively clear that the *total mass M* of the fluid should be conserved provided the velocity field **u** is always tangential to the kinematic boundary $\partial\Omega$, in particular, if **u** satisfies the no-slip boundary condition (1.41).

The continuity equation (1.34) integrated over Ω yields

$$M \equiv \int_\Omega \varrho(t, \mathbf{x}) \, d\mathbf{x} \text{ is independent of } t \in [0, T]. \tag{3.14}$$

In other words, the *total mass M* of the fluid is indeed a constant of motion.

Estimate (3.14), together with (3.12), implies that ϱ is bounded in the Lebesgue space $L^\infty(0, T; L^1(\Omega))$ by the total mass M.

Relation (3.14) is robust in the sense that it holds for any "reasonable" distributional (weak) solution of equation (1.34). An interesting counter-example will be discussed in Section 4.1.

3.3 Energy estimates

Similarly to relation (3.14), energy estimates follow directly from the (physical) fact that the total amount of certain quantities like mass or energy is either constant or at least non-increasing in time.

As we have seen in Chapter 1, the *balance of total energy* for the Navier–Stokes system (1.34)–(1.36) reads

$$\partial_t \left[\varrho\left(\tfrac{1}{2}|\mathbf{u}|^2 + P_e(\varrho) + Q(\vartheta)\right) \right] + \mathrm{div}_x \left[\left(\varrho\left(\tfrac{1}{2}|\mathbf{u}|^2 + P_e(\varrho) + Q(\vartheta)\right) + p\right)\mathbf{u} \right] + \mathrm{div}_x \mathbf{q}$$

$$= \mathrm{div}_x(\mathbb{S}\mathbf{u}) + \varrho\mathbf{f} \cdot \mathbf{u}. \tag{3.15}$$

Since the velocity field **u** vanishes on $\partial\Omega$ together with the heat flux **q**, equation (3.15) integrated with respect to the spatial variable implies

$$\frac{d}{dt} E(t) = \int_\Omega \varrho(t)\, \mathbf{f}(t) \cdot \mathbf{u}(t) d\mathbf{x}, \tag{3.16}$$

where the *total energy E* is given by formula (3.8).

Utilizing the *total mass conservation principle* (3.14), we get

$$\frac{d}{dt}E \le B_f \int_\Omega \sqrt{\varrho}\sqrt{\varrho}|\mathbf{u}|\,dx \le \frac{B_f M}{2} + B_f \int_\Omega \frac{1}{2}\varrho|\mathbf{u}|^2\,dx, \qquad (3.17)$$

where B_f is given by (3.10).

The potential P_e is determined by the elastic pressure component p_e through formula (1.22), where p_e satisfies hypothesis (3.1). In particular, as p_e is a continuous function vanishing at zero, we have

$$\varrho \mapsto \varrho P_e(\varrho) \in C[0,\infty), \quad \lim_{\varrho \to 0+}\varrho P_e(\varrho) = 0.$$

Moreover, there is a positive constant c such that

$$\varrho P_e(\varrho) \ge c\varrho^\gamma - \frac{1}{c} \quad \text{for any } \varrho \ge 0, \qquad (3.18)$$

in particular,

$$\varrho P_e(\varrho) \ge c|p_e(\varrho)| - \frac{1}{c} \quad \text{for } \varrho \ge 0 \qquad (3.19)$$

changing the value of c if necessary.

Consequently, coming back to (3.17), we discover

$$\frac{d}{dt}E \le \frac{B_f M}{2} + \frac{|\Omega|}{c} + B_f E;$$

and another application of Gronwall's lemma yields the estimate

$$E(t) \le \left(E_0 + T\left(\frac{B_f M}{2} + \frac{|\Omega|}{c}\right)\right)\exp(tB_f) \quad \text{for all } t \in [0,T], \qquad (3.20)$$

where the initial energy E_0 is specified in (3.7).

The same inequality with

$$E = E[\varrho,\mathbf{u}] \equiv \frac{1}{2}\int_\Omega \varrho|\mathbf{u}|^2 + \varrho P(\varrho)\,dx$$

holds in the *barotropic case* where $p = p_e(\varrho)$ and the equations (1.34) and (1.35) form a closed system.

Boundedness of the total energy E entails several *a priori* estimates:

(1) the density ϱ is bounded in $L^\infty(0,T;L^\gamma(\Omega))$;
(2) the kinetic and thermal energy densities $\frac{1}{2}\varrho|\mathbf{u}|^2$ and $\varrho Q(\vartheta)$ are both bounded in the space $L^\infty(0,T;L^1(\Omega))$;
(3) the momentum $\varrho\mathbf{u}$ belongs to $L^\infty(0,T;L^{(2\gamma/\gamma+1)}(\Omega;R^N))$.

Note that the third estimate follows easily from the first and second one as

$$\varrho\mathbf{u} = \sqrt{\varrho}\,\sqrt{\varrho}\mathbf{u},$$

and Hölder's inequality (2.1) can be used to conclude. All estimates depend only on the data and T as the case may be.

In particular, the elastic pressure component $p_e(\varrho)$ is integrable as a consequence of (3.19). The fact that $p_e(\varrho)$ is dominated by the potential $\varrho P_e(\varrho)$ is intimately related to the so-called Δ_2-*condition*, well known in the theory of Orlicz spaces, which is satified by p_e. To be more specific, we report the following result.

Lemma 3.1   *Let* $p_e \in C[0, \infty)$ *be a continuous function, non-decreasing on an interval* $[\underline{\varrho}, \infty)$, *and satisfying the* Δ_2-*condition:*

$$p_e(2\varrho) \le c_2 p_e(\varrho) \text{ for all } \varrho \ge \underline{\varrho}.$$

Then there exists a constant $c > 0$ *such that*

$$p_e(\varrho) \le c(1 + \varrho P_e(\varrho)) \text{ for all } \varrho \ge 0. \tag{3.21}$$

Proof   As both $p_e(\varrho)$ and $\varrho P_e(\varrho)$ are bounded on the interval $[0, 2\underline{\varrho}]$, it is enough to show (3.21) for $\varrho \ge 2\underline{\varrho}$.

We have

$$\varrho P_e(\varrho) = \varrho \int_1^\varrho \frac{p_e(z)}{z^2} \, dz = \varrho \left(\int_1^{\frac{\varrho}{2}} \frac{p_e(z)}{z^2} \, dz + \int_{\frac{\varrho}{2}}^\varrho \frac{p_e(z)}{z^2} \, dz \right)$$

$$\ge \varrho \left(\int_1^{\frac{\varrho}{2}} \frac{p_e(z)}{z^2} \, dz + \frac{p_e\left(\frac{\varrho}{2}\right)}{\varrho} \right) \ge c p_e(\varrho) - \frac{1}{c}$$

for a certain constant $c > 0$.

$\square$

3.4   Viscous dissipation

Equations (1.35) and (1.36) are "parabolic" as they both contain a second order spatial operator applied to $\mathbf{u}, \vartheta$ respectively. These terms represent the *viscous dissipation* of the mechanical energy into heat. The best way to illuminate this phenomenon is to replace the thermal energy equation (1.36) by the entropy equation (1.26). Note that (1.26) can be deduced from (1.36) and (1.34) provided all terms in these equations are regular.

The entropy equation integrated over Ω gives rise to the integral identity

$$\int_0^\tau \int_\Omega \frac{\mathbb{S} : \nabla_x \mathbf{u}}{\vartheta} + \frac{\kappa(\vartheta)|\nabla_x \vartheta|^2}{\vartheta^2} dx \, dt$$

$$= \int_\Omega (\varrho s)(\tau) \, dx - \int_\Omega (\varrho s)(0) dx \quad \text{for any } \tau \in [0, T]. \tag{3.22}$$

It follows from hypotheses (3.1) and (3.2) that the density dependent part of the entropy (cf. (3.9)) is dominated by the "elastic" part of the internal energy:

$$|\varrho P_\vartheta(\varrho)| \le c(1 + \varrho P_e(\varrho)) \quad \text{for a certain } c > 0. \tag{3.23}$$

Moreover, we have

$$\varrho \int_1^\vartheta \frac{c_v(z)}{z} \, \mathrm{d}z \le \varrho Q(\vartheta) \quad \text{for all } \vartheta > 0, \ \varrho \ge 0 \tag{3.24}$$

as

$$\int_1^\vartheta \frac{c_v(z)}{z} \, \mathrm{d}z \le 0 \quad \text{for } 0 < \vartheta \le 1$$

and

$$\int_1^\vartheta \frac{c_v(z)}{z} \, \mathrm{d}z \le \int_1^\vartheta c_v(z) \, \mathrm{d}z = Q(\vartheta) - Q(1).$$

Consequently, by virtue of the energy estimates (3.20), relation (3.22) yields

$$\int_0^T \int_\Omega \frac{\mathbb{S} : \nabla_x \mathbf{u}}{\vartheta} + \frac{\kappa(\vartheta)|\nabla_x \vartheta|^2}{\vartheta^2} \mathrm{d}\mathbf{x} \, \mathrm{d}t - \operatorname{ess\,inf}_{t\in[0,T]} \int_\Omega \varrho(t) \log(\vartheta(t)) \mathrm{d}\mathbf{x}$$

$$\le c(E_0, B_f, T) - \int_\Omega (\varrho s)(0) \mathrm{d}\mathbf{x}. \tag{3.25}$$

Now we can use (3.25) together with hypothesis (3.3) to discover the estimates

$$\operatorname{ess\,sup}_{t\in[0,T]} \int_\Omega \varrho(t) |\log(\vartheta(t))| \, \mathrm{d}\mathbf{x} \tag{3.26}$$

$$+ \int_0^T \int_\Omega |\nabla_x \vartheta^{\frac{\alpha}{2}}|^2 + |\nabla_x \log(\vartheta)|^2 \mathrm{d}\mathbf{x} \, \mathrm{d}t \le c(E_0, S_0, B_f, T),$$

where S_0 is the lower bound on the total entropy at $t = 0$ (see (3.7)).

At this stage we shall need the following auxilliary result.

Lemma 3.2 *Let $v \in W^{1,2}(\Omega)$, and let ϱ be a non-negative function such that*

$$0 < M \le \int_\Omega \varrho \, \mathrm{d}\mathbf{x}, \int_\Omega \varrho^\gamma \, \mathrm{d}\mathbf{x} \le E_0, \tag{3.27}$$

where $\Omega \subset R^N$ is a bounded domain and $\gamma > 1$.

Then there exists a constant c depending solely on M, E_0 such that

$$\|v\|_{L^2(\Omega)}^2 \le c(E_0, M) \left(\|\nabla_x v\|_{L^2(\Omega)}^2 + \left(\int_\Omega \varrho |v| \mathrm{d}\mathbf{x} \right)^2 \right).$$

Proof Were the conclusion of Lemma 3.2 not true, there would be a sequence $\{\varrho_n\}_{n=1}^\infty$ of non-negative functions satisfying (3.27) and $\{v_n\}_{n=1}^\infty \subset W^{1,2}(\Omega)$

such that

$$\|v_n\|_{L^2(\Omega)}^2 \geq c_n\left(\|\nabla_x v_n\|_{L^2(\Omega)}^2 + \left(\int_\Omega \varrho_n|v_n|\,\mathrm{d}\mathbf{x}\right)^2\right) \quad \text{with } c_n \to \infty. \quad (3.28)$$

Setting $w_n = v_n\|v_n\|_{L^2(\Omega)}^{-1}$ we have

$$w_n \to w \text{ in } L^q(B) \quad \text{for any compact } B \subset \Omega \text{ and any } 2 \leq q < 2^*, \quad (3.29)$$

where 2^* is the *critical Sobolev exponent*,

$$2^* = \frac{2N}{N-2} \quad \text{for } N \geq 3, \; 2^* \text{ arbitrary finite if } N = 1,2. \quad (3.30)$$

Here, we have used compactness of the imbedding $W^{1,2}(O) \subset L^q(O)$, where O is a regular (Lipschitz) bounded domain (see Theorem 2.7).

Moreover, by virtue of (3.28), w is constant, specifically,

$$w = \frac{1}{\sqrt{|\Omega|}}. \quad (3.31)$$

Next, in accordance with hypothesis (3.27), there exists $k = k(M, E_0)$ large enough that

$$T_k(\varrho_n) \to \overline{T_k(\varrho)} \text{ weakly in } L^\beta(\Omega) \text{ for any finite } \beta \geq 1,$$

where

$$\int_\Omega \overline{T_k(\varrho)}\,\mathrm{d}\mathbf{x} \geq \frac{M}{2}, \quad (3.32)$$

and T_k are the cut-off functions introduced in (2.11), (2.12).

On the other hand, (3.29) implies

$$0 = \lim_{n\to\infty}\int_B \varrho_n|w_n|\,\mathrm{d}\mathbf{x} \geq \lim_{n\to\infty}\int_B T_k(\varrho_n)|w_n|\,\mathrm{d}\mathbf{x}$$

$$= \int_B \overline{T_k(\varrho)}|w|\,\mathrm{d}\mathbf{x} = \frac{1}{|\Omega|}\int_B \overline{T_k(\varrho)}\,\mathrm{d}\mathbf{x} \quad \text{for any compact } B \subset \Omega$$

in contrast with (3.32).

$$\square$$

Combining the conclusion of Lemma 3.2 with the gradient estimates (3.26) and the energy inequality (3.20) we get

$$\vartheta^{\frac{\alpha}{2}} \text{ is bounded in the space } L^2(0,T;W^{1,2}(\Omega)), \quad (3.33)$$

and

$$\log(\vartheta) \text{ bounded in } L^2((0,T)\times\Omega) \quad (3.34)$$

by a constant depending only on the data and T. In particular, estimate (3.34) can be understood as "weak positivity" of the temperature ϑ.

Now, we can integrate (1.36) to obtain

$$\int_0^T \int_\Omega \mathbb{S} : \nabla_x \mathbf{u} \, dx \, dt \leq \int_0^T \int_\Omega \vartheta p_\vartheta(\varrho) |\text{div}_x \mathbf{u}| \, dx \, dt + c(E_0, B_{\mathsf{f}}, T). \qquad (3.35)$$

Seeing that, by virtue of Hölder's inequality (2.1),

$$\|\vartheta \, p_\vartheta(\varrho)\|_{L^2(\Omega)} \leq \|\vartheta\|_{L^{\frac{2N}{N-2}}(\Omega)} \|p_\vartheta(\varrho)\|_{L^N(\Omega)}, \quad N > 2,$$

one can use the estimates (3.20), (3.33) together with hypothesis (3.2) and the imbedding Theorem 2.7 to conclude that

$$\int_0^T \int_\Omega |\vartheta \, p_\vartheta(\varrho)|^2 \, dx \, dt \leq c(E_0, S_0, M, B_{\mathsf{f}}, T). \qquad (3.36)$$

The same estimate holds for $N = 2$ thanks to hypothesis (3.2).

In accordance with hypothesis (1.38), the relations (3.35), (3.36) give rise to the estimate

$$\|\nabla_x \mathbf{u}\|_{L^2((0,T)\times\Omega)} \leq c(E_0, S_0, B_{\mathsf{f}}, T). \qquad (3.37)$$

Note that

$$\int_\Omega \mathbb{S} : \nabla_x \mathbf{u} \, dx = \int_\Omega \mu |\nabla_x \mathbf{u}|^2 + (\lambda + \mu) |\text{div}_x \mathbf{u}|^2 dx \qquad (3.38)$$

provided $\mathbf{u}$ complies with the no-slip boundary conditions (1.41).

Now, we can apply Lemma 3.2 to conclude

$$\|\mathbf{u}\|_{L^2(0,T;W^{1,2}(\Omega;R^N))} \leq c(E_0, S_0, B_{\mathsf{f}}, T). \qquad (3.39)$$

As a matter of fact, we do not need to use Lemma 3.2 here as the velocity $\mathbf{u}$ satisfies the no-slip boundary conditions (1.41) and, consequently, estimate (3.39) follows directly from (3.37) and Poincare's inequality. Note also that estimate (3.39) is much simpler to derive in the *barotropic case*, where the total energy satisfies (1.29).

An interesting situation arises when $\mathbf{f}$ is a gradient of a scalar potential F independent of t, that is,

$$\mathbf{f} = \mathbf{f}(\mathbf{x}) = \nabla_x F(\mathbf{x}).$$

If this is the case, the contribution of the volumic forces to the total energy can be written in the form

$$\varrho \mathbf{u} \cdot \mathbf{f} = \varrho \mathbf{u} \cdot \nabla_x F = \partial_t(\varrho F) + \text{div}_x(\varrho F \mathbf{u}).$$

In other words, the term $-\varrho F$ can be included in the potential energy, and the quantity

$$\int_\Omega E - \varrho F \, dx$$

represents a Lyapunov function for the system (1.34)–(1.36).

A similar situation occurs when we assume the presence of long range inter-action forces as gravity or *Coulomb forces* in the fluid. The driving force is a gradient of a potential F, where

$$\Delta F = \varrho \text{ or, in general, } F = \mathcal{B}\varrho,$$

$\mathcal{B}$ a (non-local) self-adjoint operator. Here, we have denoted by $\Delta \equiv \text{div}_x \nabla_x$ the standard *Laplacian*.

Accordingly, one gets

$$\varrho \nabla_x F \cdot \mathbf{u} = \partial_t(\varrho F) + \text{div}_x(\varrho F \mathbf{u}) - \mathcal{B}[F]\partial_t F,$$

where, after integration over the spatial domain Ω, the last term yields another contribution to the total energy

$$\int_\Omega \mathcal{B}[F]\partial_t F \, d\mathbf{x} = \frac{d}{dt} \frac{1}{2} \int_\Omega |\sqrt{\mathcal{B}}[F]|^2 d\mathbf{x}.$$

The last part of this section will be devoted to some improvements of the *a priori* estimates on the temperature ϑ. Note that

$$\mathbf{q} = -\Delta \mathcal{K}(\vartheta),$$

with a potential $\mathcal{K}$,

$$\mathcal{K}(\vartheta) \equiv \int_0^\vartheta \kappa(z) \, dz;$$

whereas the bound (3.33) is not sufficient for the term $\mathcal{K}(\vartheta)$ to be at least integrable. The problem, of course, is integrability with respect to time.

As for regular solutions the temperature is always positive, we are allowed to multiply the thermal energy equation (1.36) by $\vartheta^{-\omega}, 0 < \omega \leq 1$. By parts integration yields

$$\omega \int_0^T \int_\Omega \frac{\kappa(\vartheta)|\nabla_x \vartheta|^2}{\vartheta^{\omega+1}} \, d\mathbf{x} \, dt$$

$$= \left[\int_\Omega \varrho H_\omega(\vartheta) \, d\mathbf{x} \right]_{t=0}^{t=T} - \int_0^T \int_\Omega \frac{1}{\vartheta^\omega} \mathbb{S} : \nabla_x \mathbf{u} + \vartheta^{1-\omega} p_\vartheta(\varrho) \text{div}_x \mathbf{u} \, d\mathbf{x} \, dt;$$

whence, in accordance with hypothesis (3.3)

$$\int_0^T \int_\Omega |\nabla_x \vartheta^{\frac{\alpha+1-\omega}{2}}|^2 \, d\mathbf{x} \, dt \leq c(E_0, S_0, B_f, T) \quad \text{for any } \omega > 0. \tag{3.40}$$

Here, we have denoted H_ω the primitive of $c_v(\vartheta)\vartheta^{-\omega}$.

Estimate (3.40) alone is still not sufficient to handle the term $\mathcal{K}(\vartheta)$. However, since we already know that the thermal energy density $\varrho Q(\vartheta)$ is bounded in $L^\infty(0, T; L^1(\Omega))$, interpolation arguments can be used to show that $\mathcal{K}(\vartheta)$ belongs to a Lebesgue space L^r with $r > 1$ at least on the regions where the density ϱ is bounded below away from zero. This will be done in Chapter 5.

3.5 *A priori* estimates—summary

The results achieved in this chapter may be summarized as follows:

Theorem 3.1 *Let $\Omega \subset R^N$, $N \geq 2$, be a bounded domain of class $C^{2+\nu}$, $\nu > 0$. Assume that the nonlinear functions p_e, p_ϑ, κ, and Q satisfy the hypotheses (3.1), (3.2), (3.3), and (3.5) with*

$$\gamma > \frac{N}{2}. \tag{3.41}$$

Let $\varrho, \mathbf{u}$, and ϑ be a regular (smooth) solution of the Navier–Stokes system (1.34)–(1.36) on $(0,T) \times \Omega$, where $\mathbf{u}$ and ϑ satisfy the boundary conditions (1.41) and (1.42). Moreover, let ϱ, ϑ be strictly positive at time $t = 0$, and let (3.7) hold.
Then the density ϱ is a positive function,

$$\varrho \text{ bounded in } L^\infty(0,T; L^\gamma(\Omega)); \tag{3.42}$$

the momentum

$$\varrho\mathbf{u} \text{ is bounded in } L^\infty(0,T; L^{m_\infty}(\Omega; R^N)) \cap L^2(0,T; L^{m_2}(\Omega; R^N)); \tag{3.43}$$

the convective term

$$\varrho\mathbf{u} \otimes \mathbf{u} \text{ is bounded in } L^2(0,T; L^{c_2}(\Omega; R^{N^2})), \tag{3.44}$$

and the velocity

$$\mathbf{u} \text{ is bounded in } L^2(0,T; W_0^{1,2}(\Omega; R^N)), \tag{3.45}$$

where the exponents m_∞, m_2, and c_2 are given through

$$\left.\begin{array}{l} m_\infty = \frac{2\gamma}{\gamma+1} \\[2mm] m_2 = \frac{2N\gamma}{2N+\gamma(N-2)} \end{array}\right\} > \frac{2N}{N+2}, \tag{3.46}$$

$$c_2 = \frac{2N\gamma}{N+2\gamma(N-1)} > 1. \tag{3.47}$$

The temperature ϑ is a positive function, more specifically, we have

$$\log(\vartheta) \text{ bounded in } L^2((0,T) \times \Omega), \tag{3.48}$$

$$\vartheta^{\frac{\alpha+1-\omega}{2}} \text{ bounded in } L^2(0,T; W^{1,2}(\Omega)) \quad \text{for any } 0 < \omega < 1; \tag{3.49}$$

and

$$Q(\vartheta) \text{ is bounded in } L^\infty(0,T; L^1(\Omega)). \tag{3.50}$$

The norm of these quantities in the respective function spaces is bounded by a constant $c = c(E_0, S_0, B_f, T)$ depending only on the total energy E_0 and the total entropy S_0 evaluated at instant $t = 0$, the amplitude B_f of the driving force $\mathbf{f}$, and on T.

All *a priori* estimates claimed in Theorem 3.1 were proved earlier in this section. The values of the costants m_∞, m_2, and c_2 are easily obtained with the help of Hölder's inequality and the imbedding

$$W^{1,2}(\Omega) \subset L^{2^*}(\Omega), \quad 2^* = \frac{2N}{N-2} \text{ for } N > 2, 2^* \text{ arbitrary finite if } N = 2$$

$$(3.51)$$

(cf. Theorem 2.7). Indeed the value m_∞ was computed in Section 3.3. As for m_2, we have

$$\varrho \text{ bounded in } L^\infty(0, T; L^\gamma(\Omega)), \mathbf{u} \text{ bounded in } L^2(0, T; L^{2^*}(\Omega; R^N)),$$

and the desired estimate follows by a direct application of Hölder's inequality (2.1). The value of c_2 can be obtained in a similar way.

3.6 Bibliographical notes

3.1 *A priori* estimates on the density ϱ represent one of the major stumbling blocks in developing a mathematical theory for the Navier–Stokes system (1.34)–(1.36). As we have seen in Chapter 1, the density ϱ can be computed through the system of equations (1.1) provided the velocity $\mathbf{u}$ is regular. In particular, uniform *a priori* bounds on the divergence of the velocity field are necessary to obtain uniform bounds, both from below and from above, on ϱ. Thus we end up in a closed circle of arguments as uniform bounds on $\text{div}_x \mathbf{u}$ are needed to find bounds on the density ϱ but those are not available from the standard energy estimates presented in Section 3.3. As suggested by Choe and Jin (see Theorems 1.3, 1.4 of [21]), the following three issues are intimately interrelated:

- uniform upper bounds on the density ϱ;
- uniform boundedness of ϱ below away from zero;
- uniform bounds on $\text{div}_x \mathbf{u}$.

As for the question of boundedness of $\text{div}_x \mathbf{u}$, one should keep in mind that, since the celebrated work of Leray ([78]), generations of mathematicians have been trying in vain to obtain uniform bounds on $\nabla_x \mathbf{u}$ for the *incompressible Navier–Stokes equations* in three space dimensions. The same problem for compressible fluids might be even more delicate...

3.2 In some problems, it is more convenient to assume that the total mass M is infinite. A standard hypothesis for problems posed on *unbounded domains* is $\varrho \to \varrho_\infty > 0$ for large values of $\mathbf{x}$ (see e.g. [63], [64], [92] etc.) Clearly, the purpose of such an assumption is to avoid difficulties with vacuum zones or regions of very low density which would necessarily appear for large values of $\mathbf{x}$.

On the other hand, there is an interesting result of Xin [124] concerning the Navier–Stokes system for a compressible fluid occupying the whole space R^3 with the initial density ϱ_0 compactly supported. It is shown that there is no global in-time regular solution of the problem. We remark in this regard that the Navier–Stokes system is a model of a non-dilute fluid, in which the density should be bounded below away from zero. It is natural, therefore, to expect the initial-value problem to be ill posed when vacuum zones are allowed to appear at the initial time.

3.3 The energy estimates are entirely standard and correspond to those for the incompressible Navier–Stokes equations. In general, boundedness of the total energy is a property that is expected to hold for any physically grounded problem. On the other hand, it is also well known that this type of estimate is usually not sufficient to construct solutions via compactness arguments. *A priori estimates* arising from energy balance equations are crucial when solving nonlinear equations. A systematic study of *a priori* estimates in connection with the so-called weak (or variational) solutions of evolutionary problems can be traced back to the classical monographs of Ladyzhenskaya *et al.* [75], J.-L. Lions [81], Friedman [56], and many others.

Even though it might seem essential that the elastic pressure potential $\varrho P_{\mathrm{e}}(\varrho)$ dominates the pressure term $p_e(\varrho)$ itself (see (3.19)), it is not the case. Note that a constitutive equation like $p(\varrho) \approx \exp(\varrho)$ violates the Δ_2 condition stated in Lemma 3.1 and, consequently, it does not satisfy (3.19). On the other hand, as we shall see in Chapter 4 below, there are refined *pressure estimates* which ensure that the elastic part p_e belongs to some Orlicz space built up on the space–time domain $(0, T) \times \Omega$ whose topology is stronger than that of L^1. In particular, in accordance with Theorem 2.10, bounded sequences in this space are equi-integrable and, consequently, weakly precompact in $L^1((0, T) \times \Omega)$.

3.4 It is well known that dissipative terms are usually a source of strong *a priori* estimates. The solutions of infinite-dimensional dissipative systems represented by "parabolic equations" enjoy very high regularity, and they might even be analytic at any positive time regardless of the smoothness of initial data as is the case for the standard "heat equation" (see e.g. the monograph [75] of Ladyzhenskaya *et al.*).

On the other hand, as already pointed out, the equations (1.35), (1.36) are degenerate parabolic because of the presence of the density ϱ in the leading term, which may vanish on some non-void part of the domain Ω. As a matter of fact, we are not able to recover any of the regularizing properties that (1.35), (1.36) might share with their parabolic counterparts. Similarly to the incompressible Navier–Stokes system in three space dimensions, it is still an oustanding open problem whether or not the solutions of (1.34)–(1.36) emanating from smooth initial data rest smooth on any (large) time interval. The only information resulting from the presence of dissipative terms is thus square integrability of the velocity gradient $\nabla_x \mathbf{u}$, and similar estimates of certain power functions of the temperature ϑ.

4

VARIATIONAL SOLUTIONS

Strictly speaking, there is no reason for quantities like the density ϱ or the temperature ϑ to be continuous or even differentiable with respect to the spatial variable $\mathbf{x}$. Their physical interpretation as the specific mass and "specific thermal energy" respectively makes it conceivable to consider ϱ and ϑ to be bounded (non-negative) measures which may be absolutely continuous with respect to the standard Lebesgue measure on R^N (see e.g. [116]). If the *velocity* field $\mathbf{u}$ is regular, that is, if the motion is smooth, the continuity equation (1.34) still makes sense through formula (1.4). In particular, if the fluid occupies a bounded domain Ω on the boundary of which the normal component of the velocity vanishes, one would expect the continuity equation to hold in some sense provided both ϱ and $\mathbf{u}$ were extended to be zero outside Ω.

One way to accommodate possible discontinuities is to consider a *weak formulation* of the problem, where derivatives of the unknown functions are understood in the sense of *distributions* (see Section 2.1). The common philosophy of this approach is to replace pointwise values of the state variables by their *spatial averages*. Intuitively one cannot expect to "measure" a value of a physically observable quantity at a point but only its integral averages over a (possibly small) neighborhood of this point.

On the other hand, these integral averages should be continuous in time in order to characterize the state of the system. Note that this approach is, in fact, much closer to the physical nature of the problem because it makes use of the basic balance laws in their natural integral form. As we have already seen in Chapter 2, the instantaneous values of integrable functions can be defined through their spatial averages as in Definition 2.1.

4.1 The equation of continuity

4.1.1 Weak formulation

The *principle of mass conservation* can be expressed by means of the family of integral equations (1.4), which are nothing other than a system of differential equations governing the time evolution of the mass in a given spatial region B. Note that (1.4) makes sense when ϱ and $\mathbf{u}$ are merely continuous functions. In order to deduce the weak formulation, we shall use the following *change of variables formula*.

Theorem 4.1 [42, Chapter 3.4, Section 3.4.3, Theorem 2] *Let $\eta\colon R^N \to R$ be a Lipschitz function, and $v \in L^1(R^N)$.*
Then for a.a. $r \in R$ we have

$$v|_{\{\eta=r\}} \text{ integrable in the sense}$$

$$\text{of the } (N-1)\text{-dimensional Hausdorff measure } \mathrm{d}\sigma,$$

and

$$\int_{R^N} v(\mathbf{x})|\nabla_x\eta(\mathbf{x})|\,\mathrm{d}\mathbf{x} = \int_R \left[\int_{\{\eta=r\}} v(\mathbf{x})\,\mathrm{d}\sigma\right]\mathrm{d}r.$$

Now, let $\eta \in \mathcal{D}(\Omega)$ be a non-negative function. By virtue of the Morse–Sard theorem, the *upper level sets* $\{\eta > r\}$,

$$\{\eta > r\} \equiv \{\mathbf{x} \in \Omega | \eta(\mathbf{x}) > r\},$$

are of class C^∞ for a.a. $r \in [0, \infty)$. Consequently, we can take $B = \{\eta > r\}$ in (1.4) and use Theorem 4.1 for $v = \varrho\mathbf{u} \cdot (\nabla_x\eta/|\nabla_x\eta|)$ to obtain

$$\frac{\mathrm{d}}{\mathrm{d}t}\int_0^\infty \left[\int_{\{\eta>r\}} \varrho(t,\mathbf{x})\,\mathrm{d}\mathbf{x}\right]\mathrm{d}r = \int_\Omega \varrho(t,\mathbf{x})\mathbf{u}(t,\mathbf{x}) \cdot \nabla_x\eta(\mathbf{x})\,\mathrm{d}\mathbf{x}.$$

Seeing that

$$\int_0^\infty \left[\int_{\{\eta>r\}} \varrho(t,\mathbf{x})\,\mathrm{d}\mathbf{x}\right]\mathrm{d}r = \int_0^\infty \int_\Omega \mathrm{sgn}^+[\eta(\mathbf{x}) - r]\varrho(t,\mathbf{x})\,\mathrm{d}\mathbf{x}\,\mathrm{d}r$$

$$= \int_\Omega \varrho(t,\mathbf{x})\eta(\mathbf{x})\,\mathrm{d}\mathbf{x}$$

we can rewrite (1.4) in the form:

$$\frac{\mathrm{d}}{\mathrm{d}t}\int_{R^N} \varrho(t,\mathbf{x})\eta(\mathbf{x})\,\mathrm{d}\mathbf{x} = \int_{R^N} \varrho(t,\mathbf{x})\mathbf{u}(t,\mathbf{x}) \cdot \nabla_x\eta(\mathbf{x})\,\mathrm{d}\mathbf{x} \qquad (4.1)$$

for any non-negative $\eta \in \mathcal{D}(\Omega)$. This equation describes the time evoution of the *spatial averages*

$$t \in (0,T) \mapsto \int_{R^N} \varrho(t,\mathbf{x})\eta(\mathbf{x})\,\mathrm{d}\mathbf{x}, \quad \eta \in \mathcal{D}(\Omega).$$

It is easy to see that (4.1) is in fact equivalent to an infinite family of integral identities

$$\int_R \int_{R^N} \varrho\partial_t\varphi + \varrho\mathbf{u} \cdot \nabla\varphi\,\mathrm{d}\mathbf{x}\,\mathrm{d}t = 0 \qquad (4.2)$$

which are to hold for any *test function* $\varphi \in \mathcal{D}((0,T) \times \Omega)$. If (4.2) holds, we shall say that equation (1.34) is satisfied in the sense of distributions on $(0,T) \times \Omega$, or, shortly, in $\mathcal{D}'((0,T) \times \Omega)$.

The class of test functions considered in (4.2) is inconveniently small in view of future applications. However, a simple density argument which may be justified via the smoothing operators $v \mapsto [v]_y^\varepsilon$ introduced (2.3)–(2.4) shows that the integral identity (4.2) remains valid for a much larger family of functions:

Lemma 4.1 *Let*

$$\varrho \in L^\infty(0,T;L^\gamma(\Omega)), \quad \varrho\mathbf{u} \in L^\infty(0,T;L^{m_\infty}(\Omega;R^N)) \tag{4.3}$$

satisfy the continuity equation (1.34) *in* $\mathcal{D}'((0,T)\times\Omega)$, *that is, identity* (4.2) *holds for any* $\varphi \in \mathcal{D}((0,T)\times\Omega)$.

Then (4.2) *holds for any test function* φ *such that*

$$\varphi \in L^1((0,T)\times\Omega), \quad \mathrm{supp}[\varphi] \subset O \subset (0,T)\times\Omega,$$

$$\partial_t\varphi \in L^1(0,T;L^{\gamma'}(\Omega)), \quad \nabla_x\varphi \in L^1(0,T;L^{m'_\infty}(\Omega;R^N)),$$

where O *is compact, and*

$$\frac{1}{\gamma}+\frac{1}{\gamma'}=1, \quad \frac{1}{m_\infty}+\frac{1}{m'_\infty}=1.$$

Remark In accordance with the *a priori* estimates stated in Theorem 3.1, one can anticipate that a "reasonable" solution will satisfy hypothesis (4.3) for $m_\infty = 2\gamma/(\gamma+1)$.

Remark Here, the *support of a distribution* $\mathrm{supp}[v]$ is a closed set determined as the complement of the open set

$$\{\mathbf{y} \in O \,|\, \text{there exists a neighborhood } U(\mathbf{y}), v|_U = 0 \text{ in } \mathcal{D}'(U)\}.$$

Remark If Ω is smooth, the test functions φ belonging to the class

$$\varphi \in L^1((0,T)\times\Omega), \partial_t\varphi \in L^1(0,T;L^{\gamma'}(\Omega)), \nabla_x\varphi \in L^1(0,T;L^{m'_\infty}(\Omega;R^N))$$

have a well-defined trace on the boundary $\partial((0,T)\times\Omega)$, and the conclusion of Lemma 4.1 remains valid if the hypothesis $\mathrm{supp}[\varphi] \subset O$ is replaced by

$$\varphi|_{\partial((0,T)\times\Omega)} = 0.$$

4.1.2 Average continuity of the density

The time evolution of the *integral averages*

$$t \in (0,T) \mapsto \int_{R^N} \varrho(t,\mathbf{x})\eta(\mathbf{x})\,\mathrm{d}\mathbf{x}$$

is governed by equation (4.1). In particular, these mappings, considered for a fixed η as functions of t only, are (globally) Lipschitz continuous on the time interval $(0,T)$ for any $\eta \in \mathcal{D}(\Omega)$ provided ϱ, $\mathbf{u}$ belong to the regularity class (4.3).

Since the set of smooth functions $\varphi \in \mathcal{D}(\Omega)$ is dense in $L^{\gamma'}(\Omega)$ (cf. Theorem 2.2), the continuity of the integral averages can be rephrased as

$$\varrho \in C([0,T]; L^{\gamma}_{\text{weak}}(\Omega)). \tag{4.4}$$

More precisely, the instantaneous value $\varrho(t)$ introduced in Definition 2.1 is representable by a function $\varrho(t) \in L^{\gamma}(\Omega)$ for *any* $t \in (0,T)$. Similarly, $\varrho(0+) \in L^{\gamma}(\Omega)$, $\varrho(T-) \in L^{\gamma}(\Omega)$, and the function

$$t \mapsto \begin{cases} \int_{R^N} \varrho(0+)\eta \, dx & \text{for } t = 0, \\[2mm] \int_{R^N} \varrho(t)\eta \, dx & \text{if } t \in (0,T), \\[2mm] \int_{R^N} \varrho(T-)\eta \, dx & \text{for } t = T \end{cases}$$

is continuous on $[0,T]$ for any $\eta \in L^{\gamma'}(\Omega)$.

Now, it is easy to see that any weak solution ϱ, $\mathbf{u}$ belonging to class (4.3) will satisfy the integral identity

$$\int_0^T \int_\Omega \varrho(t,\mathbf{x})\partial_t\varphi(t,\mathbf{x}) + (\varrho\mathbf{u})(t,\mathbf{x}) \cdot \nabla\varphi(t,\mathbf{x}) \, dx \, dt$$

$$= \int_\Omega \varrho(T,\mathbf{x})\varphi(T,\mathbf{x}) \, dx - \int_\Omega \varrho(0,\mathbf{x})\varphi(0,\mathbf{x}) \, dx \tag{4.5}$$

for any test function φ,

$$\varphi, \partial_t\varphi \in L^1(0,T; L^{\gamma'}(\Omega)), \quad \varphi \in L^1(0,T; W_0^{1,m'_\infty}(\Omega)). \tag{4.6}$$

4.1.3 Total mass conservation revisited

As we have seen in Section 3.2, the total mass M of any *classical solution* ϱ of the continuity equation (1.34) is a conserved quantity provided the velocity $\mathbf{u}$ is tangential to the kinematic boundary $\partial\Omega$ at any instant t. Our aim now is to examine to what extent this property remains valid in the class of weak solutions introduced above.

To this end, consider a velocity field $\mathbf{u}$,

$$\mathbf{u} \in L^2(0,T; W_0^{1,2}(\Omega; R^N)),$$

and a non-negative density function ϱ,

$$\varrho \in L^\infty(0,T; L^{\gamma}(\Omega)),$$

satisfying the integral identity (4.2) for any $\varphi \in \mathcal{D}((0,T) \times \Omega)$.

In order to understand all possible difficulties that may occur, let us start with a simple example when $N = 1$, and, accordingly, we can set $\Omega = (-1,1)$.

Consider the velocity

$$u = u(x) = (1-x)^\alpha (1+x)^\alpha, \quad x \in (-1,1), \tag{4.7}$$

which obviously belongs to the class $L^2(I; W_0^{1,2}(-1,1))$ provided $\alpha > \frac{1}{2}$.

Now, the density ϱ can be taken in the form

$$\varrho(t,x) = H\left(\int_{-1}^{x} \frac{1}{u(s)} \, ds - t\right) \frac{1}{u(x)}, \quad t \in R, \ x \in (-1,1), \tag{4.8}$$

where $H \in C^1(R)$ is an arbitrary function.

It is easy to check that ϱ, u defined by (4.7), (4.8) solve the continuity equation (1.34) on $I \times (-1,1)$ for any open interval $I \subset R$. Thus a suitable choice of H gives rise to a non-trivial solution ϱ emanating from the zero initial state

$$\varrho(0,x) = \varrho_0(x) = 0 \quad \text{for all } x \in (-1,1).$$

In particular, such a solution apparently violates the *principle of total mass conservation* (3.14). Note that ϱ, u are smooth inside $(0,T) \times \Omega$ and represent a classical solution of (1.34). On the other hand, the density ϱ is singular at the boundary $\partial\Omega$, and this singularity is "much stronger" than the rate of decay to zero of the velocity $\mathbf{u}$ in the same region. If the total mass conservation principle is to be preserved, these two phenomena must be in balance; in particular, we must be able to control the boundary behavior of the density.

We shall need the following assertion, which is a straightforward consequence of *Hardy's inequality*.

Theorem 4.2 [104, Theorem 21.5] *Let $\Omega \subset R^N$ be a bounded Lipschitz domain. Let $v \in W_0^{1,p}(\Omega)$ with $1 \le p < \infty$.*

Then the function

$$\frac{v}{\mathrm{dist}(\cdot, \partial\Omega)} \ \textit{belongs to } L^p(\Omega),$$

and

$$\left\| \frac{v}{\mathrm{dist}(\cdot, \partial\Omega)} \right\|_{L^p(\Omega)} \le c(p,\Omega) \|v\|_{W_0^{1,p}(\Omega)}.$$

Remark Here, the symbol $\mathrm{dist}(\mathbf{x}, K)$ denotes the (non-negative) *distance function*,

$$\mathrm{dist}(\mathbf{x}, K) = \min_{\mathbf{y} \in K} |\mathbf{x} - \mathbf{y}|, \quad K \subset R^N \text{ a non-void compact set.}$$

Note that $\mathrm{dist}(\mathbf{x}, K)$ is Lipschitz with constant 1 (see e.g. Chapter 1 in [127]).

With help of Theorem 4.2, one can identify the class of weak solutions for which the total mass M is conserved.

Proposition 4.1   *Let $\Omega \subset R^N$ be a bounded Lipschitz domain. Let*

$$\varrho \in L^2(0,T;L^2(\Omega)), \quad \mathbf{u} \in L^2(0,T;W_0^{1,2}(\Omega;R^N)) \tag{4.9}$$

solve the continuity equation (1.34) in $\mathcal{D}'((0,T) \times \Omega)$.

   Then the functions ϱ, $\mathbf{u}$ solve (1.34) in $\mathcal{D}'((0,T) \times R^N)$ provided they were extended to be zero outside Ω.

   If, in addition, $\varrho \in L^\infty(0,T;L^\gamma(\Omega))$ with $\gamma > 1$, then the total mass

$$M = \int_\Omega \varrho(t,\mathbf{x})\,\mathrm{d}\mathbf{x} \text{ is independent of } t \in [0,T].$$

Remark   In accordance with (4.4), we have $\varrho \in C([0,T];L_{\text{weak}}^\gamma(\Omega))$ and, consequently, the total mass $M = M(t)$ is well defined for any $t \in [0,T]$.

Proof   Consider a sequence

$$\chi_n(y) = \chi(ny), \quad y \in R^1, \tag{4.10}$$

where

$$\chi \in C^\infty(R), \quad \chi(y) = 0 \text{ if } y \le \tfrac{1}{4}, \quad \chi(y) = 1 \text{ if } y \ge \tfrac{3}{4}, \quad \chi \text{ non-decreasing on } R.$$

   By virtue of Lemma 4.1, we have

$$\int_R \int_{R^N} \varrho \partial_t \varphi + \varrho \mathbf{u} \cdot \nabla \varphi \,\mathrm{d}\mathbf{x}\,\mathrm{d}t$$

$$= \int_R \int_\Omega \varrho (1 - \chi_n(\text{dist}[\cdot,\partial\Omega]))\partial_t\varphi + \varrho \mathbf{u} \cdot \nabla_x \left((1 - \chi_n(\text{dist}[\cdot,\partial\Omega]))\varphi\right)\mathrm{d}\mathbf{x}\,\mathrm{d}t$$

for any $\varphi \in \mathcal{D}((0,T) \times R^N)$.

   Now,

$$\int_R \int_\Omega \varrho\,(1 - \chi_n(\text{dist}[\cdot,\partial\Omega]))\,\mathrm{d}\mathbf{x}\,\mathrm{d}t \to 0 \quad \text{as } n \to \infty,$$

$$\int_R \int_\Omega \varrho\,(1 - \chi_n(\text{dist}[\cdot,\partial\Omega]))\,\mathbf{u} \cdot \nabla_x\varphi\,\mathrm{d}\mathbf{x}\,\mathrm{d}t \to 0 \quad \text{as } n \to \infty$$

by virtue of Lebesgue's theorem, while

$$\left| \int_R \int_\Omega \varphi \varrho \mathbf{u} \cdot \nabla_x \chi_n(\text{dist}[\cdot,\partial\Omega])\,\mathrm{d}\mathbf{x}\,\mathrm{d}t \right|$$

$$\le c(\chi,\Omega) \int_R \int_{\{\text{dist}[\mathbf{x},\partial\Omega] \le \frac{1}{n}\}} |\varphi| \varrho \frac{|\mathbf{u}|}{\text{dist}[\cdot,\partial\Omega]}\,\mathrm{d}\mathbf{x}\,\mathrm{d}t,$$

where the integral on the right-hand side tends to zero by virtue of hypothesis (4.9), Hölder's inequality (2.1), Theorem 4.2, and Lebesgue's theorem.

$\square$

4.1.4   Weak sequential stability

Unlike their classical counterparts, the weak solutions to (1.34) form a compact set with respect to the topologies induced by the *a priori* estimates obtained in Chapter 3. This property is essential for the existence theory developed later in this book.

Lemma 4.2   *Let Ω be an arbitrary domain in R^N.*
     Assume that ϱ_n, $\mathbf{u}_n$ solve the continuity equation (1.34) in $\mathcal{D}'((0,T)\times\Omega)$, and

$$\varrho_n \to \varrho \ weakly(\text{-}^*) \ in \ L^\infty(0,T;L^\gamma(\Omega)),$$

$$\mathbf{u}_n \to \mathbf{u} \ weakly \ in \ L^2(0,T;W^{1,2}(\Omega;R^N)),$$

where

$$\gamma > \frac{2N}{N+2}. \tag{4.11}$$

Moreover, let the sequence

$$\{\varrho_n \mathbf{u}_n\}_{n=1}^\infty \ be \ bounded \ in \ L^\infty(0,T;L^{m_\infty}(\Omega;R^N))$$

for a certain $m_\infty > 1$.
     Then

$$\varrho_n \mathbf{u}_n \to \varrho\mathbf{u} \ weakly(\text{-}^*) \ in \ L^\infty(0,T;L^{m_\infty}(\Omega;R^N)), \tag{4.12}$$

where ϱ, $\mathbf{u}$ solve (1.34) in $\mathcal{D}'((0,T)\times\Omega)$. If, in addition, ϱ_n are non-negative functions, then so is ϱ.

Proof   Clearly, it is enough to show (4.12), more precisely, we have to prove that any possible accummulation point of the sequence $\{\varrho_n\mathbf{u}_n\}_{n=1}^\infty$ coincides with $\varrho\mathbf{u}$.

Let $\eta \in \mathcal{D}(\Omega)$. Since ϱ_n, $\mathbf{u}_n$ satisfy (1.34) in $\mathcal{D}'((0,T)\times\Omega)$, we deduce

$$\frac{\mathrm{d}}{\mathrm{d}t}\int_{R^N}\varrho_n(t)\eta\,\mathrm{d}\mathbf{x} = \int_{R^N}\varrho_n(t)\mathbf{u}_n(t)\cdot\nabla_x\eta\,\mathrm{d}\mathbf{x} \quad in \ \mathcal{D}'(0,T),$$

where, in accordance with Theorem 2.6 and Hölder's inequality (2.1), the right-hand side is bounded in $L^2(0,T)$.

Consequently, we can use Corollary 2.1 to conclude that

$$\varrho_n \to \varrho \ in \ C([0,T];L^\gamma_{\mathrm{weak}}(\Omega)).$$

By virtue of Theorem 2.8, the space $L^\gamma(K)$ is compactly imbedded in $W^{-1,2}(K)$ for any bounded (smooth) domain $K \subset \Omega$; whence

$$\varrho_n \to \varrho \ in \ C([0,T];W^{-1,2}(K)).$$

Seeing that

$\varphi \mathbf{u}_n \to \eta \mathbf{u}$ weakly in $L^2(0, T; W_0^{1,2}(\Omega; R^N))$ for any $\varphi \in \mathcal{D}((0, T) \times \Omega)$,

we have

$$\int_0^T \int_\Omega \varphi \varrho_n \mathbf{u}_n \, d\mathbf{x} \, dt \to \int_0^T \int_\Omega \varphi \varrho \mathbf{u} \, d\mathbf{x} \, dt$$

which yields (4.12).

$\square$

4.1.5 Renormalized solutions

It is well known that hyperbolic systems of conservation laws are not well posed in the class of distributional solutions. The appearance of discontinuities (shocks) represents another source of energy dissipation not captured by the weak formulation which is perfectly time reversible. In general, an additional family of so-called entropy inequalities must be added to identify the physically admissible solution. For a hyperbolic equation like (1.34), which is *linear* in ϱ, the concept of entropy solutions was accordingly modified by DiPerna and Lions [32]. The resulting class of solutions bears the name *renormalized* as this procedure can be considered as a simple *rescaling* of the original state variables.

Multiplying the continuity equation (1.34) by expression $B'(\varrho)$, where B is a smooth function, we formally deduce the equation

$$\partial_t B(\varrho) + \mathrm{div}_x(B(\varrho)\mathbf{u}) + b(\varrho)\mathrm{div}_x\mathbf{u} = 0, \tag{4.13}$$

where

$$b(z) = B'(z)z - B(z). \tag{4.14}$$

Such a procedure can be perfectly justified for classical solutions of the problem. At the level of distributional solutions, however, (4.13) may represent an additional constraint, which is, in general, not captured by the weak formulation (4.2).

Definition 4.1 We shall say that ϱ (and $\mathbf{u}$) is a *renormalized solution* of the *continuity equation* (1.34) on $(0, T) \times \Omega$ if (4.13) holds in $\mathcal{D}'((0, T) \times \Omega)$ for any functions

$$B \in C[0, \infty) \cap C^1(0, \infty), \quad b \in C[0, \infty) \text{ bounded on } [0, \infty),$$

$$B(0) = b(0) = 0 \tag{4.15}$$

satisfying (4.14) for all $z > 0$.

Remark It is easy to see that any renormalized solution of (1.34) in the sense of Definition 4.1 satisfies (1.34) in $\mathcal{D}'((0,T) \times \Omega)$ as one can take $B(z) \equiv z$, $b(z) \equiv 0$.

Resolving (4.14) with respect to B we obtain

$$B(\varrho) = \varrho \int_1^\varrho \frac{b(z)}{z^2}\,\mathrm{d}z + B(1)\varrho \quad \text{for any } \varrho \geq 0, \tag{4.16}$$

which is, in fact, equivalent to (4.14).

As already pointed out in Chapter 2, nonlinear compositions do not commute, in general, with weak limits in the space $L^\infty(0,T; L^\gamma(\Omega))$. Consequently, the set of renormalized solutions need not be closed (weakly stable). However, there is a very simple criterion for a weak solution ϱ to be a renormalized one based on the following *"commutator"* lemma.

Lemma 4.3 *Let* $\Omega \subset R^N$ *be a domain. Let*

$$\varrho \in L^p(\Omega), \quad \mathbf{u} \in W^{1,q}(\Omega; R^N)$$

be given functions with $q \geq 1$ finite,

$$\frac{1}{p} + \frac{1}{q} \leq 1. \tag{4.17}$$

Then

$$\|[\mathrm{div}_x(\varrho\mathbf{u})]_x^\varepsilon - \mathrm{div}_x([\varrho]_x^\varepsilon \mathbf{u})\|_{L^1(K)} \leq c(K)\|\varrho\|_{L^p(\Omega)}\|\mathbf{u}\|_{W^{1,q}(\Omega;R^N)}, \tag{4.18}$$

and

$$[\mathrm{div}_x(\varrho\mathbf{u})]_x^\varepsilon - \mathrm{div}_x([\varrho]_x^\varepsilon \mathbf{u}) \to 0 \text{ in } L^1(K) \quad \text{as } \varepsilon \to 0 \tag{4.19}$$

for any compact $K \subset \Omega$. Here $v \mapsto [v]_x^\varepsilon$ are the smoothing (convolution) operators introduced in (2.4).

Proof To begin with, observe that the quantity

$$[\mathrm{div}_x(\varrho\mathbf{u})]_x^\varepsilon - \mathrm{div}_x([\varrho]_x^\varepsilon \mathbf{u})$$

is well defined on K provided $\varepsilon > 0$ is small enough. Moreover, as (4.19) obviously holds for smooth functions ϱ which are dense in $L^1(K)$, one can use the Banach–Steinhaus theorem to observe that it is enough to show the bound (4.18).

To this end, we write

$$[\operatorname{div}_x(\varrho\mathbf{u})]_x^\varepsilon(\mathbf{x}) - \operatorname{div}_x([\varrho]_x^\varepsilon\mathbf{u})(\mathbf{x})$$

$$= \int_{R^N} \varrho(\mathbf{y})(\mathbf{u}(\mathbf{x}) - \mathbf{u}(\mathbf{y})) \cdot \nabla_x\theta^\varepsilon(|\mathbf{x}-\mathbf{y}|)\,\mathrm{d}\mathbf{y} + [\varrho]_x^\varepsilon\operatorname{div}_x\mathbf{u}(\mathbf{x}) \qquad \text{for any } \mathbf{x} \in K,$$

where the first term can be expressed as follows:

$$\int_{R^N} \varrho(\mathbf{y})(\mathbf{u}(\mathbf{x}) - \mathbf{u}(\mathbf{y})) \cdot \nabla_x\theta^\varepsilon(|\mathbf{x}-\mathbf{y}|)\,\mathrm{d}\mathbf{y}$$

$$= \int_{R^N} \varrho(\mathbf{x}-\mathbf{z})\frac{\mathbf{u}(\mathbf{x}) - \mathbf{u}(\mathbf{x}-\mathbf{z})}{|\mathbf{z}|} \cdot \nabla_x\theta^\varepsilon(|\mathbf{z}|)|\mathbf{z}|\,\mathrm{d}\mathbf{z}.$$

By virtue of hypothesis (4.17), Theorem 2.9, and Hölder's inequality (2.1) the right-hand side of the above identity is bounded in $L^1(K)$ as claimed in (4.18).
□

The following assertion is a straightforward consequence of Lemma 4.3.

Proposition 4.2 *Let $\Omega \subset R^N$ be an arbitrary domain. Let*

$$r \in L^2(0,T;L^2(\Omega)), \quad \mathbf{u} \in L^2(0,T;W^{1,2}(\Omega;R^N)), \quad \text{and} \quad h \in L^1(0,T;L^1(\Omega))$$

satisfy

$$\partial_t r + \operatorname{div}_x(r\mathbf{u}) = h \qquad \text{in } \mathcal{D}'((0,T)\times\Omega). \tag{4.20}$$

Then we have

$$\partial_t B(r) + \operatorname{div}_x(B(r)\mathbf{u}) + b(r)\operatorname{div}_x\mathbf{u} = B'(r)h \qquad \text{in } \mathcal{D}'((0,T)\times\Omega) \tag{4.21}$$

for any bounded $b \in C^1[0,\infty)$ and B given by (4.16).

Proof Applying the regularizing operators $v \mapsto [v]_x^\varepsilon$ to both sides of (4.20) we obtain

$$\partial_t[r]_x^\varepsilon + \nabla_x[r]_x^\varepsilon \cdot \mathbf{u} + [r]_x^\varepsilon\operatorname{div}_x\mathbf{u} = [h]_x^\varepsilon + s^\varepsilon \quad \text{a.a.} \quad \text{on } O \tag{4.22}$$

for any bounded open set $O \subset \overline{O} \subset (0,T)\times\Omega$ provided $\varepsilon > 0$ is small enough. Here, "applying" means using the kernels θ_x^ε defined by (2.3) as test functions in the weak formulation of (4.20).

By virtue of Lemma 4.3, we get

$$s^\varepsilon \in L^1(O), \quad s^\varepsilon \to 0 \text{ in } L^1(O) \quad \text{as } \varepsilon \to 0.$$

Observe that, by virtue of our hypotheses, we have at least

$$\partial_t[r]_x^\varepsilon \in L^1(O), \quad \nabla_x[r]_x^\varepsilon \in L^2(O)$$

so that (4.22) makes sense.

Equation (4.22) multiplied on $B'(r)$, where B is determined through (4.16) with $b \in \mathcal{D}(0, \infty)$, gives rise to identity

$$\partial_t B([r]_x^\varepsilon) + \text{div}_x(B([r]_x^\varepsilon)\mathbf{u}) + b([r]_x^\varepsilon)\text{div}_x\mathbf{u} = B'([r]_x^\varepsilon)(h + s^\varepsilon);$$

which, in the limit $\varepsilon \to 0$, yields (4.21). Here, we have used the fact that

$$\partial_y B(v) = B'(v)\partial_y v \text{ a.a. on } O, \quad \text{for } y = t, x_i, \ i = 1, \ldots, N,$$

for any Lipschitz B and any $v \in W^{1,p}(O)$, $p \geq 1$ (see Theorem 2.1.11 in [127]).

$\square$

Corollary 4.1 *Let $\Omega \subset R^N$ be an arbitrary domain. Let*

$$\varrho \in L^2(0, T; L^2(\Omega))$$

solve the continuity equation (1.34) in $\mathcal{D}'((0, T) \times \Omega)$ with

$$\mathbf{u} \in L^2(0, T; W^{1,2}(\Omega; R^N)).$$

Then ϱ is a renormalized solution of (1.34) on $(0, T) \times \Omega$.

Proof Take a sequence $b_n \in \mathcal{D}(0, \infty)$ which is uniformly bounded and converges to b uniformly on compacts in $[0, \infty)$. Use Proposition 4.2 together with Lebesgue's convergence theorem.

$\square$

Keeping in mind our convention on *instantaneous values* (cf. Definition 2.1 of Chapter 2), any composed function $B(\varrho)$, where ϱ is a renormalized solution of (1.34), can be understood as continuous in time with respect to the weak L^γ-topology:

$$B(\varrho) \in C([0, T]; L^\gamma_{\text{weak}}(\Omega))$$

provided $\varrho \in L^\infty(0, T; L^\gamma(\Omega))$ with $\gamma > 1$. By the same token,

$$\varrho \in C([0, T]; L^\gamma_{\text{weak}}(\Omega)),$$

which leads to a natural question, namely, whether or not the identity

$$B(\varrho)(t) = B(\varrho(t)) \tag{4.23}$$

holds for *any* $t \in [0, T]$. Here, $B(\varrho)(t)$ is the instantaneous value of the function $B(\varrho)$ at t in the sense of Definition 2.1 while $B(\varrho(t))$ stands for the composition of the function $B : R \to R$ applied to the instantaneous value of ϱ at t. Even though (4.23) holds for a.a. $t \in (0, T)$, there is *a priori* no reason for (4.23) to be satisfied identically on $[0, T]$.

The way out of this rather ambiguous situation is provided by the following result which states that any renormalized solution ϱ of (1.34) is, in fact, strongly continuous as a function of time with values in $L^1(\Omega)$, in particular, (4.23) holds for any $t \in [0, T]$.

Proposition 4.3 *Let $\Omega \subset R^N$ be a bounded domain. Let $\varrho \geq 0$,*

$$\varrho \in L^\infty(0,T;L^\gamma(\Omega)), \mathbf{u} \in L^2(0,T;W^{1,2}(\Omega;R^N)), \gamma > \frac{2N}{N+2},$$

be a renormalized solution of the continuity equation (1.34).
    Then

$$\varrho \in C([0,T];L^1(\Omega)).$$

Proof Since ϱ is a renormalized solution, we have

$$\partial_t T_k(\varrho) + \mathrm{div}_x(T_k(\varrho)\mathbf{u}) + (T_k'(\varrho)\varrho - T_k(\varrho))\,\mathrm{div}_x\mathbf{u} = 0 \quad \text{in } \mathcal{D}'((0,T) \times \Omega),$$
$$(4.24)$$

where T_k are the cut-off functions defined in (2.11). In particular,

$$T_k(\varrho) \text{ belong to } C([0,T];L^\gamma_{\text{weak}}(\Omega)) \quad \text{for any } k \geq 1.$$

Similarly to the proof of Proposition 4.2, we can apply the smoothing operators to both sides of (4.24) to obtain

$$\partial_t [T_k(\varrho)]^\varepsilon_x + \mathrm{div}_x\left([T_k(\varrho)]^\varepsilon_x\mathbf{u}\right) = s^\varepsilon \text{ a.a.} \quad \text{on } (0,T) \times K, K \subset \Omega \text{ compact},$$
$$(4.25)$$

where, by virtue of Lemma 4.3, the quantities s^ε are bounded in the space $L^2(0,T;L^1(K))$ independently of ε.
    Multiplying (4.25) by $2[T_k(\varrho)]^\varepsilon_x$ we get

$$\partial_t([T_k(\varrho)]^\varepsilon_x)^2 + \mathrm{div}_x\left(([T_k(\varrho)]^\varepsilon_x)^2\mathbf{u}\right) + ([T_k(\varrho)]^\varepsilon_x)^2\,\mathrm{div}_x\mathbf{u} = 2[T_k(\varrho)]^\varepsilon_x s^\varepsilon.$$

Thus for any fixed $\eta \in \mathcal{D}(\Omega)$, the family of functions

$$t \mapsto \int_\Omega ([T_k(\varrho)]^\varepsilon_x)^2(t,\mathbf{x})\eta(\mathbf{x})\,\mathrm{d}\mathbf{x}, \varepsilon > 0 \text{ is precompact in } C[0,T].$$

On the other hand,

$$[T_k(\varrho)]^\varepsilon_x(t,\cdot) \to T_k(\varrho)(t,\cdot) \text{ in } L^2(\Omega) \text{ as } \varepsilon \to 0 \quad \text{for any } t \in [0,T],$$

where, of course, $T_k(\varrho)(t)$ are represented by their instantaneous values. Consequently, the function

$$t \mapsto \int_\Omega (T_k(\varrho))^2(t,\mathbf{x})\eta(\mathbf{x})\,\mathrm{d}\mathbf{x} \text{ belongs to } C[0,T]$$

for any fixed $\eta \in \mathcal{D}(\Omega)$. This implies

$$T_k(\varrho) \in C([0,T];L^2(\Omega)) \quad \text{for any fixed } k \geq 1$$

(see Theorem 2.11).
    As $\gamma > 1$ and $k \geq 1$ were arbitrary, we conclude $\varrho \in C([0,T];L^1(\Omega))$.

$\square$

4.1.6 *Variational solutions*

Motivated by the previous discussion we introduce the concept of a *variational solution of the continuity equation* (1.34) supplemented with the no-slip boundary conditions (1.41) for the velocity **u**.

Definition 4.2 Let $\Omega \subset R^N$, $N \geq 2$, be a bounded domain. We shall say that a pair of functions ϱ, **u** is a *variational solution* of the equation of continuity (1.34) with **u** satisfying the boundary conditions (1.41) on $(0, T) \times \Omega$ if

(1) the density ϱ is a non-negative function,

$$\varrho \in L^\infty(0, T; L^\gamma(\Omega)), \quad \gamma > \frac{2N}{N+2};$$

(2) the velocity **u** belongs to the space $L^2(0, T; W_0^{1,2}(\Omega; R^N))$,

$$\varrho\mathbf{u} \in L^\infty(0, T; L^{m_\infty}(\Omega; R^N)) \quad \text{for a certain } m_\infty > 1;$$

(3) the functions ϱ, **u** represent a renormalized solution of equation (1.34) in the sense of Definition 4.1 on the whole space $(0, T) \times R^N$ provided ϱ, **u** were extended to be zero outside Ω.

Note that, in accordance with Proposition 4.3, the density component ϱ of any variational solution belongs to the space

$$C([0, T]; L^1(\Omega)) \cap C([0, T]; L^\gamma_{\text{weak}}(\Omega)).$$

Moreover, the *total mass*

$$M \equiv \int_\Omega \varrho(t, \mathbf{x}) \, d\mathbf{x} \text{ is independent of } t \in [0, T].$$

4.2 Momentum equation

4.2.1 *Weak formulation*

In a similar way to that in the preceding section, one can use the integral version of the *momentum equation* (1.8) to obtain

$$\frac{d}{dt} \int_{R^N} \varrho(t, \mathbf{x})\mathbf{u}(t, \mathbf{x}) \cdot \eta(\mathbf{x}) \, d\mathbf{x}$$

$$= \int_{R^N} (\varrho\mathbf{u} \otimes \mathbf{u})(t, \mathbf{x}) : \nabla_x\eta(\mathbf{x}) + p(t, \mathbf{x})\text{div}_x\eta(\mathbf{x}) - \mathbb{S}(t, \mathbf{x}) : \nabla_x\eta(\mathbf{x}) \, d\mathbf{x}$$

$$+ \int_{R^N} \varrho(t, \mathbf{x})\mathbf{f}(t, \mathbf{x}) \cdot \eta(\mathbf{x}) \, d\mathbf{x} \tag{4.26}$$

governing the time evolution of the spatial averages

$$t \in (0, T) \mapsto \int_{R^N} (\varrho \mathbf{u})(t, \mathbf{x}) \cdot \eta(\mathbf{x}) \, d\mathbf{x} \quad \text{for } \eta \in [\mathcal{D}(\Omega)]^N.$$

Equivalently, we have

$$\int_R \int_{R^N} \varrho \mathbf{u} \cdot \partial_t \varphi + (\varrho \mathbf{u} \otimes \mathbf{u}) : \nabla_x \varphi + p \operatorname{div}_x \varphi \, d\mathbf{x} \, dt$$

$$= \int_R \int_{R^N} \mathbb{S} : \nabla_x \varphi - \varrho \mathbf{f} \cdot \varphi \, d\mathbf{x} \, dt \quad \text{for any } \varphi \in [\mathcal{D}((0, T) \times \Omega)]^N, \quad (4.27)$$

which is a *weak formulation of the momentum equation* (1.35).

Analogously as in Section 4.1, it is easy to show that the system of admissible test functions can be enlarged as follows.

Lemma 4.4 *Let* $\Omega \subset R^N$ *be an arbitrary open set. Let*

$$\varrho \in L^\infty(0, T; L^\gamma(\Omega)), \quad \gamma > \frac{N}{2},$$

$$\varrho \mathbf{u} \in L^\infty(0, T; L^{m_\infty}(\Omega; R^N)) \cap L^2(0, T; L^{m_2}(\Omega; R^N)), \quad m_\infty, m_2 > \frac{2N}{N+2},$$

$$\varrho \mathbf{u} \otimes \mathbf{u} \in L^2(0, T; L^{c_2}(\Omega; R^{N^2})), \quad c_2 > 1,$$

$$p \in L^1(0, T; L^1(\Omega)), \quad \text{and} \quad \mathbb{S} \in L^2(0, T; L^2(\Omega; R^{N^2}))$$

solve the momentum equation (1.35) in $\mathcal{D}'((0, T) \times \Omega)$.

Then the integral identity (4.27) holds for any test function

$$\varphi \in L^1(0, T; L^{\gamma'}(\Omega; R^N))$$

such that

$$\operatorname{supp}[\varphi] \subset O \subset (0, T) \times \Omega, \quad O \subset R^{N+1} \text{ compact,}$$

$$\partial_t \varphi \in L^2(0, T; L^{m_2'}(\Omega; R^N)), \quad \nabla_x \varphi \in L^2(0, T; L^{c_2'}(\Omega; R^{N^2})),$$

and

$$\operatorname{div}_x \varphi \in L^\infty((0, T) \times \Omega), \quad \nabla_x \varphi \in L^2(0, T; L^2(\Omega; R^{N^2})),$$

where

$$\frac{1}{\gamma} + \frac{1}{\gamma'} = \frac{1}{m_2} + \frac{1}{m_2'} = \frac{1}{c_2} + \frac{1}{c_2'} = 1.$$

Remark Similarly to Lemma 4.1, compactness of $\operatorname{supp}[\varphi]$ can be replaced by a more general condition that φ vanishes on the boundary $\partial((0, T) \times \Omega)$.

4.2.2 Average continuity

The time evolution of the integral averages

$$t \in (0,T) \mapsto \int_{\Omega} (\varrho \mathbf{u})(t,\mathbf{x}) \cdot \eta(\mathbf{x})\,d\mathbf{x}$$

is governed by (4.26), in particular, under the hypotheses of Lemma 4.4, these are (Hölder) continuous functions of $t \in [0,T]$. As $\varrho\mathbf{u}$ belongs to the space $L^{\infty}(0,T;L^{m_{\infty}}(\Omega;R^{N}))$, we infer

$$\varrho\mathbf{u} \in C([0,T];L^{m_{\infty}}_{\text{weak}}(\Omega;R^{N})).$$

More precisely, the *instantaneous values* (see Definition 2.1) of $(\varrho\mathbf{u})(0+)$, $(\varrho\mathbf{u})(t)$, $t \in (0,T)$, and $(\varrho\mathbf{u})(T-)$ belong to the Lebesgue space $L^{m_{\infty}}(\Omega)$, and the function

$$t \mapsto \begin{cases} \int_{R^{N}} (\varrho\mathbf{u})(0+) \cdot \eta\,d\mathbf{x} & \text{for } t = 0, \\ \int_{R^{N}} (\varrho\mathbf{u})(t) \cdot \eta\,d\mathbf{x} & \text{if } t \in (0,T), \\ \int_{R^{N}} (\varrho\mathbf{u})(T-) \cdot \eta\,d\mathbf{x} & \text{for } t = T \end{cases}$$

is continuous on $[0,T]$ for any $\eta \in [L^{m'_{\infty}}(\Omega)]^{N}$.

Unfortunately, there is no information available concerning the time continuity of the velocity $\mathbf{u}$ itself, not even in the sense of spatial averages. This is obviously due to possible appearance of vacuum regions—sets where the density ϱ vanishes. No matter how contradictory this might seem, there are no *a priori* estimates available to prevent this from happening except for $N = 1$ (see also Section 3.1 of Chapter 3 and [66]).

As we have seen, the density $\varrho(t)$ and the momentum $\varrho\mathbf{u}(t)$ possess a specific instantaneous value at any $t \in [0,T]$. We shall show that these values comply with a seemingly obvious *compatibility condition* $\varrho\mathbf{u}(t) = 0$ a.a. on the set $\{\varrho(t) = 0\}$.

Lemma 4.5 *Let $\Omega \subset R^{N}$ be an open set. Assume*

$$\varrho \in C([0,T];L^{\gamma}_{\text{weak}}(\Omega)), \quad \mathbf{u} \in L^{2}(0,T;L^{2}(\Omega,R^{N})),$$

and

$$\varrho\mathbf{u} \in C([0,T];L^{m_{\infty}}_{\text{weak}}(\Omega;R^{N})), \quad \varrho|\mathbf{u}|^{2} \in L^{\infty}(0,T;L^{1}(\Omega))$$

for certain $\gamma > 1$, $m_{\infty} > 1$.

Then

$$\varrho\mathbf{u}(t,\cdot) = 0 \quad a.a. \quad \text{on the set } \{\mathbf{x} \in \Omega \mid \varrho(t,\mathbf{x}) = 0\}$$

for any $t \in [0,T]$.

Proof Let $B \subset R^N$ be a bounded ball, $\tau \in [0, T]$. Set

$$B_0 = B \cap \{\mathbf{x} \in R^N \mid \varrho(\tau, \mathbf{x}) = 0\}.$$

Since the norm in L^1 is weakly lower semi-continuous, we get

$$\int_{B_0} |\varrho\mathbf{u}(\tau)| \, d\mathbf{x} \leq \liminf_{t \to \tau, t \in (0,T)} \int_{B_0} |\varrho\mathbf{u}(t)| \, d\mathbf{x}$$

$$\leq \liminf_{t \to \tau, t \in (0,T)} \left(\int_{B_0} \varrho(t) \, d\mathbf{x} \right)^{1/2} \left(\operatorname*{ess\,sup}_{t \in (0,T)} \int_{\Omega} \varrho|\mathbf{u}|^2(t) \, d\mathbf{x} \right)^{1/2},$$

where we have used the Cauchy–Schwartz inequality (the inequality (2.1) for $p_1 = p_2 = 2$).

As ϱ is weakly continuous in t, we conclude that

$$\lim_{t \to \tau, t \in (0,T)} \int_{B_0} \varrho(t) \, d\mathbf{x} = 0,$$

which completes the proof.

$\square$

In accordance with the common philosophy of Chapter 1, the momentum $\varrho\mathbf{u}$ is more convenient to play the role of a state variable than the velocity $\mathbf{u}$ as it possesses instantaneous values at any $t \in [0, T]$ which are continuous functions of t. Accordingly, the initial conditions (1.43) are formulated in terms of $\varrho\mathbf{u}$ rather than $\mathbf{u}$. Note, however, that both formulations are equivalent provided the density ϱ is strictly positive.

4.2.3 Weak sequential stability

The set of all weak solutions to the momentum equation (1.35) is weakly compact with respect to the topologies related to *a priori* estimates obtained in Chapter 3. To see this, we employ once more the arguments already used in the proof of Lemma 4.2:

Lemma 4.6 *Let $\Omega \subset R^N$ be an arbitrary domain. Let*

$$\left. \begin{array}{c} \varrho_n\mathbf{u}_n \to \varrho\mathbf{u} \text{ weakly(-*) in } L^\infty(0, T; L^{m_\infty}(\Omega)), \\[2mm] \mathbf{u}_n \to \mathbf{u} \text{ weakly in } L^2(0, T; W^{1,2}(\Omega; R^N)), \\[2mm] p_n \to p \text{ weakly in } L^1((0, T) \times \Omega), \\[2mm] p_n \text{ bounded in } L^\infty(0, T; L^1(\Omega)) \oplus L^2((0, T) \times \Omega), \\[2mm] \mathbb{S}_n \to \mathbb{S} \text{ weakly in } L^2(0, T; L^2(\Omega; R^{N^2})). \end{array} \right\} \qquad (4.28)$$

and

$$\mathbf{f}_n \to \mathbf{f} \quad \text{in } L^1((0, T) \times \Omega),$$

where ϱ_n, $\mathbf{u}_n$, p_n, $\mathbb{S}_n$ and $\mathbf{f}_n$ satisfy the momentum equation (1.35) in $D'((0,T) \times \Omega)$. Let, moreover,

$$m_\infty > \frac{2N}{N+2},$$

and let $\mathbf{f}_n$ be bounded a.a. on the set $(0,T) \times \Omega$ by a constant B_f as in (3.10).

Then the limit functions ϱ, $\mathbf{u}$, p, $\mathbb{S}$, and $\mathbf{f}$ satisfy the same system (1.35) in $\mathcal{D}'((0,T) \times \Omega)$.

Proof   Obviously, it is enough to show

$$\varrho_n \mathbf{u}_n \otimes \mathbf{u}_n \to \varrho \mathbf{u} \otimes \mathbf{u} \text{ weakly in } L^2(0,T; L^{c_2}(K; R^{N^2})) \qquad (4.29)$$

for any bounded $K \subset \Omega$, where

$$\frac{1}{c_2} = \frac{1}{m_\infty} + \frac{1}{2^*} \leq 1,$$

$$2^* \text{ arbitrary finite for } N = 2, \ 2^* = \frac{2N}{N-2} \quad \text{for } N \geq 3.$$

By virtue of the imbedding Theorem 2.6 and Hölder's inequality (2.1), the quantities $\varrho_n \mathbf{u}_n \otimes \mathbf{u}_n$, $n = 1,2,\ldots$ are bounded in $L^2(0,T; L^{c_2}(K; R^{N^2}))$. Moreover, since the time derivative $\partial_t \varrho_n \mathbf{u}_n$ can be computed by means of (4.26), we have

$$\varrho_n \mathbf{u}_n \to \varrho \mathbf{u} \quad \text{in } C([0,T]; L^{m_\infty}_{\text{weak}}(K, R^N)) \qquad (4.30)$$

for any bounded set $K \subset \Omega$.

Finally, as

$$m_\infty > \frac{2N}{N+2}$$

we can use Theorem 2.8 to conclude

$$\varrho_n \mathbf{u}_n \to \varrho \mathbf{u} \quad \text{in } C([0,T]; W^{-1,2}(K; R^N))$$

for any bounded domain $K \subset \overline{K} \subset \Omega$. As

$$\varphi \mathbf{u}_n \to \varphi \mathbf{u} \text{ weakly in } L^2(0,T; W^{1,2}_0(\Omega; R^N)) \text{ for any } \varphi \in \mathcal{D}((0,T) \times \Omega),$$

(4.29) follows.

$\square$

We have shown that both the continuity equation (1.34) and the momentum equation (1.35) are *weakly sequentially stable* with respect to the topologies given by the *a priori* estimates derived in Chapter 3. A more difficult task will be to show *weak continuity* of the quantities like $\mathbb{S}$ and p which are determined by the constitutive equations (1.37) and (1.39). For Newtonian fluids, where $\mathbb{S}$ depends linearly on the velocity gradient, this does not seem to be a difficult problem unless the viscosity coefficients μ and λ are allowed to depend on ϱ or other physical quantities. On the other hand, as we have seen in Section 1.4 of Chapter 1, the pressure p is typically a *nonlinear* function of the density ϱ and the temparature ϑ. Consequently, the problem of weak continuity becomes much more delicate.

4.2.4 Energy inequality

Strong solutions considered in Chapter 1 satisfy the mechanical energy equation (1.10) which, integrated with respect to the spatial variable, gives rise to the integral identity

$$\left[\int_\Omega \tfrac{1}{2}\varrho|\mathbf{u}|^2 \,\mathrm{d}\mathbf{x} \right]_{t=0}^{t=\tau} + \int_0^\tau \int_\Omega \mathbb{S} : \nabla_x \mathbf{u} \,\mathrm{d}\mathbf{x}\,\mathrm{d}t$$

$$= \int_0^\tau \int_\Omega p \operatorname{div}_x \mathbf{u} \,\mathrm{d}\mathbf{x}\,\mathrm{d}t + \int_0^\tau \int_\Omega \varrho \mathbf{f} \cdot \mathbf{u} \,\mathrm{d}\mathbf{x}\,\mathrm{d}t$$

for any $\tau \in [0, T]$. Moreover, if $\mathbb{S}$ and p are given by the constitutive equation (1.37) and (1.39) respectively, this relation reads

$$\int_\Omega \varrho(\tau) \left(\tfrac{1}{2}|\mathbf{u}(\tau)|^2 + P_{\mathrm{e}}(\varrho(\tau)) \right) \,\mathrm{d}\mathbf{x} + \int_0^\tau \int_\Omega \mu|\nabla_x \mathbf{u}|^2 + (\lambda + \mu)|\operatorname{div}_x \mathbf{u}|^2 \,\mathrm{d}\mathbf{x}\,\mathrm{d}t$$

$$= \int_\Omega \varrho(0) \left(\tfrac{1}{2}|\mathbf{u}(0)|^2 + P_{\mathrm{e}}(\varrho(0)) \right) \,\mathrm{d}\mathbf{x} + \int_0^\tau \int_\Omega \vartheta p_\vartheta \operatorname{div}_x \mathbf{u} \,\mathrm{d}\mathbf{x}\,\mathrm{d}t$$

$$+ \int_0^\tau \int_\Omega \varrho \mathbf{f} \cdot \mathbf{u} \,\mathrm{d}\mathbf{x}\,\mathrm{d}t, \tag{4.31}$$

where the *elastic pressure potential* P_{e} is given by (1.22). Here, the term

$$\int_0^\tau \int_\Omega \mu|\nabla_x \mathbf{u}|^2 + (\lambda + \mu)|\operatorname{div}_x \mathbf{u}|^2 - \vartheta p_\vartheta \operatorname{div}_x \mathbf{u} \,\mathrm{d}\mathbf{x}\,\mathrm{d}t \tag{4.32}$$

is responsible for *dissipation* of the mechanical energy into heat.

It is an outstanding open problem if the energy equality (4.31), which has been derived under the assumption of smoothness of all quantities involved, holds for any weak (distributional) solution satisfying (1.34), (1.35) in $\mathcal{D}'((0,T) \times \Omega)$.

The point is that the dissipative term $\mathbb{S} : \nabla_x \mathbf{u}$ is known to be bounded only in the space $L^1((0,T) \times \Omega)$. Although the instantaneous value of $[\mathbb{S} : \nabla_x \mathbf{u}](t)$ introduced in Definition 2.1 is absolutely continuous with respect to the Lebesgue measure and can be computed on the basis of the velocity $\mathbf{u}(t)$ for a.a. $t \in (0,T)$, it may happen that

$$[\mathbb{S} : \nabla_x \mathbf{u}](\tau) \notin L^1(\Omega) \text{ for a certain } \tau \in [0, T].$$

If this occurs, the total mechanical energy defect is no longer captured by the integral term (4.32), and a non-negative measure must be added to the left-hand side of (4.31). Accordingly, we obtain the *mechanical energy inequality* in the form

$$E_{\text{mech}}(\tau) + \int_0^\tau \int_\Omega \mu |\nabla_x \mathbf{u}|^2 + (\lambda + \mu)|\text{div}_x \mathbf{u}|^2 \, dx \, dt$$

$$\leq E_{\text{mech}}(0) + \int_0^\tau \int_\Omega \vartheta p_\vartheta \, \text{div}_x \mathbf{u} \, dx \, dt + \int_0^\tau \int_\Omega \varrho \mathbf{f} \cdot \mathbf{u} \, dx \, dt \qquad (4.33)$$

for a.a. $\tau \in (0, T)$, where we have denoted E_{mech} the mechanical energy,

$$E_{\text{mech}}(t) = E_{\text{mech}}[\varrho, \mathbf{u}](t) \equiv \int_\Omega \varrho(t) \left(\tfrac{1}{2}|\mathbf{u}(t)|^2 + P_{\text{e}}(\varrho(t)) \right) \, dx. \qquad (4.34)$$

Although we do not know if such phenomena really occur, in particular for a dissipative system like (1.34), (1.35), the weak solutions are known to satisfy only (4.33) in place of (4.31). Note that the same problem appears even in the incompressible case in three space dimensions, where the possible instantaneous kinetic energy defects could be attributed to *turbulence* (see e.g. the survey article [8] by Bardos and Nicolaenko).

Dealing with weak solutions one must be careful when speaking about *instantaneous values* of quantities like E_{mech} at a specific time $t \in [0, T]$. Following our strategy started in Definition 2.1 we can define the instantaneous value of the quantity

$$\left[\tfrac{1}{2}\varrho|\mathbf{u}|^2 + \varrho P_{\text{e}}(\varrho) \right](\tau),$$

which is a non-negative bounded measure on $\overline{\Omega}$ for any fixed $\tau \in (0, T)$. Note that, by virtue of the energy estimates (3.20), one has

$$\left[\tfrac{1}{2}\varrho|\mathbf{u}|^2 + \varrho P_{\text{e}}(\varrho) \right] \in L^\infty(0, T; L^1(\Omega)).$$

Now, the instantaneous value of

$$E_{\text{mech}}(\tau) = E_{\text{mech}}[\varrho, \mathbf{u}](\tau)$$

can be *defined* as

$$\left[\tfrac{1}{2}\varrho|\mathbf{u}|^2 + \varrho P_{\text{e}}(\varrho) \right](\tau) \left[\overline{\Omega} \right]. \qquad (4.35)$$

On the other hand, there is a second possibility of understanding $E_{\text{mech}}(t)$ based on the fact that we already know the instantaneous values of the density $\varrho(t)$ and the momentum $\varrho\mathbf{u}(t)$, which are continuous functions of time on $[0, T]$.

Consider a function $\mathcal{E}_{\text{mech}} : R \times R^N \to (-\infty, \infty]$,

$$
\mathcal{E}_{\text{mech}}(r, \mathbf{m}) =
\begin{cases}
\dfrac{1}{2}\dfrac{|\mathbf{m}|^2}{r} + r P_{\text{e}}(r) & \text{if } r > 0, \\[2ex]
0 & \text{for } r = 0, \ \mathbf{m} = 0, \\[2ex]
\infty & \text{for } r \leq 0, \ (r, \mathbf{m}) \neq (0, 0).
\end{cases}
$$

Assume, to begin with, that the elastic pressure component $p_{\text{e}} = p_{\text{e}}(\varrho)$ is a non-decreasing function of ϱ. Accordingly, it can be checked by direct inspection that $\mathcal{E}_{\text{mech}}$ is a convex lower semi-continuous function on $R \times R^N$, which is bounded from below.

Taking Lemma 4.5 into account, we can *define* the *total mechanical energy* as

$$
E_{\text{mech}}(\tau) = E_{\text{mech}}[\varrho(\tau), (\varrho\mathbf{u})(\tau)] \equiv \int_\Omega \mathcal{E}_{\text{mech}}\left(\varrho(\tau), \varrho\mathbf{u}(\tau)\right) \, \mathrm{dx}, \quad \tau \in [0, T],
$$

$$(4.36)$$

where $\varrho(\tau)$, $(\varrho\mathbf{u})(\tau)$ are the instantaneous values in the sense of Definition 2.1. Since the instantaneous values of ϱ and $\varrho\mathbf{u}$ are continuous functions of time with respect to the weak L^1- topology on Ω, the function E_{mech} defined through (4.36) is lower semi-continuous on $[0, T]$, in particular,

$$
E_{\text{mech}}[\varrho(0), (\varrho\mathbf{u})(0)] \leq \liminf_{t \to 0+} E_{\text{mech}}[\varrho(t), (\varrho\mathbf{u})(t)] \tag{4.37}
$$

(see Corollary 2.2).

If the elastic pressure satisfies only hypothesis (3.1), that is, the elastic potential $\varrho \mapsto \varrho P_{\text{e}}(\varrho)$ is not necessarily convex, the mechanical energy defined through (4.36) is still a lower semi-continuous function of time. Indeed one can write

$$
\varrho P_{\text{e}}(\varrho) = B(\varrho) + C(\varrho),
$$

where, in accordance with (3.1), B is bounded and C convex on $[0, \infty)$. If ϱ is a renormalized solution of (1.34) in the sense of Definition 4.1, then $\varrho \in C([0, T]; L^1(\Omega))$, in particular,

$$
t \mapsto \int_\Omega B(\varrho(t)) \, \mathrm{dx} \text{ is a continuous function of } t \in [0, T].
$$

Even though the functions E_{mech} defined respectively by (4.35), (4.36) coincide almost anywhere on the time interval $(0, T)$, one should always keep in mind that their values could be different for certain concrete times $\tau \in (0, T)$.

There is one more thing to be pointed out in connection with the energy inequality (4.33), namely that the *dissipative term*

$$
\mathbb{S} : \nabla_x \mathbf{u} - \vartheta p_\vartheta \mathrm{div}_x \mathbf{u}
$$

appears also on the right-hand side of the thermal energy equation (1.36). If we are not able to guarantee the *energy equality* (4.31) in the class of variational solutions, we have to abandon the *equality sign* in (1.36) as well. Accordingly, the concept of a *variational solution* for (1.36) must be considerably modified as we will see in Section 4.3.

4.2.5 Variational solutions

Definition 4.3 Let $\Omega \subset R^N$ be a bounded domain. We shall say that a family of functions $\varrho\mathbf{u}$, $\mathbf{u}$, p, and $\mathbb{S}$ is a *variational solution of the momentum equation* (1.35) supplemented with the boundary conditions (1.41) on the set $(0, T) \times \Omega$ if

(1) the momentum $\varrho\mathbf{u}$ belongs to the class $L^\infty(0, T; L^{m_\infty}(\Omega; R^N))$, with

$$m_\infty > \frac{2N}{N+2};$$

(2) the velocity $\mathbf{u}$ satisfies

$$\mathbf{u} \in L^2(0, T; W_0^{1,2}(\Omega; R^N));$$

(3) $p \in L^1(0, T; L^1(\Omega))$, $\quad \mathbb{S} \in L^2(0, T; L^2(\Omega; R^{N^2}))$;

(4) the functions $\varrho\mathbf{u}$, $\mathbf{u}$, p, and $\mathbb{S}$ satisfy the momentum equation (1.35) in $\mathcal{D}'((0, T) \times \Omega)$.

Note that the convective term $\varrho\mathbf{u} \otimes \mathbf{u}$ belongs to the Lebesgue space $L^2(0, T; L^{c_2}(\Omega; R^{N^2}))$ for a certain $c_2 > 1$, and that

$$\varrho\mathbf{u} \in C([0, T]; L^{m_\infty}_{\text{weak}}(\Omega; R^N)).$$

4.3 Thermal energy equation

Continuing our discussion from Section 4.2 about the energy inequality (4.33) satisfied by the weak solutions of the momentum equation (1.35), we turn our attention to the thermal energy balance expressed through equation (1.36). Recall from the earlier exposition that we do not expect (1.36) to hold even in a weak sense but rather to be replaced by an inequality

$$\partial_t(\varrho Q(\vartheta)) + \text{div}_x(\varrho Q(\vartheta)\mathbf{u}) - \text{div}_x(\kappa(\vartheta)\nabla_x\vartheta) \geq \mathbb{S} : \nabla_x\mathbf{u} - \vartheta p_\vartheta \text{div}_x\mathbf{u}. \quad (4.38)$$

Accordingly, the Neumann boundary conditions (1.42) must be replaced by unilateral ones

$$\nabla_x\vartheta \cdot \mathbf{n}|_{\partial\Omega} \geq 0, \quad\quad\quad\quad (4.39)$$

while $\mathbf{u}$ still satisfies (1.41). The corresponding (unilateral) initial condition reads

$$\varrho Q(\vartheta)(0, \mathbf{x}) \geq \chi_0(\mathbf{x}) \text{ a.a. } \text{ on } \Omega. \quad\quad\quad\quad (4.40)$$

4.3.1 Weak formulation

As usual, a weak formulation of the problem (4.38)–(4.40) is represented by a family of integral inequalities

$$\int_0^T \int_\Omega \varrho Q(\vartheta) \partial_t \varphi + \varrho Q(\vartheta) \mathbf{u} \cdot \nabla_x \varphi + \mathcal{K}(\vartheta) \Delta \varphi \, d\mathbf{x} \, dt$$

$$\leq \int_0^T \int_\Omega (\vartheta p_\vartheta \operatorname{div}_x \mathbf{u} - \mathbb{S} : \nabla_x \mathbf{u}) \, \varphi \, d\mathbf{x} \, dt - \int_\Omega \chi_0 \varphi(0) \, d\mathbf{x} \qquad (4.41)$$

for any test function φ,

$$\varphi \geq 0, \quad \varphi \in W^{2,\infty}((0,T) \times \Omega), \quad \nabla_x \varphi \cdot \mathbf{n}|_{\partial\Omega} = 0, \quad \operatorname{supp}[\varphi] \subset [0,T) \times \overline{\Omega}. \quad (4.42)$$

Here, we have set

$$\mathcal{K}(\vartheta) \equiv \int_0^\vartheta \kappa(z) \, dz. \qquad (4.43)$$

Similarly to those in Sections 4.1, 4.2, the regularity hypotheses imposed on the admissible test functions φ can be relaxed:

Lemma 4.7 *Let $\Omega \subset R^N$ be a domain of class $C^{2+\nu}$, $\nu > 0$. Assume that the quantities*

$$\varrho Q(\vartheta) \in L^2(0,T; L^{q_2}(\Omega)), \quad q_2 > \frac{2N}{N+2},$$

$$\mathbf{u} \in L^2(0,T; W_0^{1,2}(\Omega; R^N)),$$

$$\mathcal{K}(\vartheta) \in L^1((0,T) \times \Omega),$$

$$\mathbb{S}, \vartheta p_\vartheta \in L^2((0,T) \times \Omega)$$

satisfy the integral inequality (4.41) for any test function φ as in (4.42).
Then (4.41) holds for any φ such that

$$\begin{cases} \varphi \in L^\infty((0,T) \times \Omega), \varphi \geq 0, \\[2mm] \partial_t \varphi \in L^2(0,T; L^{q_2'}(\Omega)), \quad \nabla_x \varphi \in L^\infty(0,T; W_0^{1,p}(\Omega)), \ p > N, \\[2mm] \Delta \varphi \in L^\infty((0,T) \times \Omega), \\[2mm] \operatorname{supp}[\varphi] \subset [0,T) \times \overline{\Omega}, \end{cases} \qquad (4.44)$$

where

$$\frac{1}{q_2} + \frac{1}{q_2'} = 1.$$

Proof   To begin with, we can set

$$\varphi(t,\cdot) = \varphi(0,\cdot) \quad \text{for } t < 0, \quad \varphi(t,\cdot) = 0 \quad \text{for } t > T,$$

where now the extended function satisfies (4.44) with the interval $(0,T)$ replaced by R.

Next, we set

$$\chi(t,\mathbf{x}) = \begin{cases} \Delta\varphi(t,\mathbf{x}) & \text{for } \mathbf{x} \in \Omega, \\[2mm] 0 & \text{if } \mathbf{x} \in R^N \setminus \Omega. \end{cases}$$

In accordance with our hypotheses, χ is a bounded measurable function on $R \times R^N$,

$$\int_{R^N} \chi(t,\mathbf{x})\,\mathrm{d}\mathbf{x} = 0 \quad \text{for any } t \in R.$$

Let us denote by $v = \Delta_N^{-1}(g)$ the unique solution of the Neumann problem

$$\Delta v = g \text{ in } \Omega, \quad \nabla_x g \cdot n|_{\partial\Omega} = 0, \quad \int_\Omega v\,\mathrm{d}\mathbf{x} = \int_\Omega g\,\mathrm{d}\mathbf{x} = 0.$$

Now, consider a test function

$$\varphi_{\varepsilon,\delta} = \left[\Delta_N^{-1}\left([\chi]_x^\delta - \int_\Omega [\chi]_x^\delta\,\mathrm{d}\mathbf{x} \right) + \int_\Omega \varphi\,\mathrm{d}\mathbf{x} \right]_t^\varepsilon, \quad \varepsilon,\delta > 0,$$

where $v \mapsto [v]_t^\varepsilon$, $v \mapsto [v]_x^\delta$ are the convolution operators introduced in (2.4). It is easy to check that $\varphi_{\varepsilon,\delta}$ satisfies (4.42) except for being non-negative on $[0,T]$.

On the other hand, there exists a compact set $O \subset (0,T) \times \overline{\Omega}$ such that

$$\mathrm{supp}[\varphi_{\varepsilon,\delta}] \subset O \quad \text{for all } \varepsilon,\delta.$$

Moreover, we have

$$\Delta_N^{-1}\left([\chi(t)]_x^\delta - \int_\Omega [\chi(t)]_x^\delta\,\mathrm{d}\mathbf{x} \right) + \int_\Omega \varphi(t)\,\mathrm{d}\mathbf{x} \to \varphi(t) \geq 0 \quad \text{in } C(\overline{\Omega}) \quad \text{for } \delta \to 0$$

for any fixed $t \in [0,T]$. Thus one can find a sequence $\{\psi_\delta\}_{\delta>0}$,

$$\psi_\delta \in C^\infty[0,T], \quad \mathrm{supp}[\psi] \subset [0,T), \quad \psi_\delta \to 0 \text{ in } C^1[0,T] \quad \text{for } \delta \to 0,$$

such that

$$\varphi_{\varepsilon,\delta} + \psi_\delta \geq 0$$

satisfies (4.42).

Consequently, we can take $\varphi_{\varepsilon,\delta} + \psi_\delta$ as a test function in (4.41) and pass to the limit, first for $\delta \to 0$, then for $\varepsilon \to 0$, to get the same inequality for φ.

$\square$

4.3.2 Weak solutions with a defect measure

It may seem that a significant piece of information is lost when replacing the thermal energy *equation* (1.36) by the family of integral *inequalities* (4.41). Following the idea of Alexandre and Villani [3], we introduce the concept of a *weak solution with defect measure* related to equation (1.36). In addition to (4.41), we require the total *energy inequality*

$$E(\tau) \equiv \int_\Omega \varrho(\tau) \left(\tfrac{1}{2} |\mathbf{u}(\tau)|^2 + P_e(\varrho(\tau)) + Q(\vartheta(\tau)) \right) \mathrm{d}\mathbf{x}$$

$$\leq E(0) + \int_0^\tau \int_\Omega \varrho \mathbf{f} \cdot \mathbf{u} \, \mathrm{d}\mathbf{x} \, \mathrm{d}t \tag{4.45}$$

to hold for a.a. $\tau \in (0, T)$.

It is easy to see that the family of integral inequalities (4.41), *together* with (4.45), represents a suitable weak formulation of equation (1.36); specifically, that (4.41), (4.45) define a solution and not an upper solution of (1.36). Indeed were the inequality (4.41) strict, we would get

$$\int_0^T \int_\Omega \varrho Q(\vartheta) \partial_t \varphi_\delta + \varrho Q(\vartheta) \mathbf{u} \cdot \nabla_x \varphi_\delta + \mathcal{K}(\vartheta) \Delta \varphi_\delta \, \mathrm{d}\mathbf{x} \, \mathrm{d}t + \delta$$

$$= \int_0^T \int_\Omega (\vartheta p_\vartheta \mathrm{div}_x \mathbf{u} - \mathbb{S} : \nabla_x \mathbf{u}) \, \varphi_\delta \, \mathrm{d}\mathbf{x} \, \mathrm{d}t - \int_\Omega \chi_0 \varphi_\delta(0) \, \mathrm{d}\mathbf{x}$$

for $\delta > 0$ and a certain test function φ_δ, where we can assume $0 \leq \varphi_\delta \leq 1$. Consequently, we have

$$\int_0^T \int_\Omega \varrho Q(\vartheta) \, \partial_t \varphi + \varrho Q(\vartheta) \mathbf{u} \cdot \nabla_x \varphi + \mathcal{K}(\vartheta) \Delta \varphi \, \mathrm{d}\mathbf{x} \, \mathrm{d}t + \delta$$

$$\leq \int_0^T \int_\Omega (\vartheta p_\vartheta \mathrm{div}_x \mathbf{u} - \mathbb{S} : \nabla_x \mathbf{u}) \, \varphi \, \mathrm{d}\mathbf{x} \, \mathrm{d}t - \int_\Omega \chi_0 \varphi(0) \, \mathrm{d}\mathbf{x}$$

for any $\varphi \geq \varphi_\delta$ satisfying (4.42). Thus we deduce a sharp inequality

$$\int_\Omega \varrho(\tau) Q(\vartheta)(\tau) \, \mathrm{d}\mathbf{x} - \int_\Omega \varrho(0) Q(\vartheta)(0) \, \mathrm{d}\mathbf{x}$$

$$> \int_0^\tau \int_\Omega \mathbb{S} : \nabla_x \mathbf{u} - \vartheta p_\vartheta \mathrm{div}_x \mathbf{u} \, \mathrm{d}\mathbf{x} \, \mathrm{d}t \tag{4.46}$$

for a.a. $\tau \geq \tau_0$ for a certain $\tau_0 \in (0, T)$.

On the other hand, *smooth solutions* of the Navier–Stokes system satisfy the mechanical *energy equality*

$$\int_\Omega \varrho(\tau) \left(\tfrac{1}{2}|\mathbf{u}(\tau)|^2 + P_e(\varrho(t))\right)\mathrm{d}\mathbf{x} - \int_\Omega \varrho(0) \left(\tfrac{1}{2}|\mathbf{u}(0)|^2 + P_e(\varrho(0))\right)\mathrm{d}\mathbf{x}$$

$$= \int_0^\tau \int_\Omega \vartheta p_\vartheta \,\mathrm{div}_x\mathbf{u} - \mathbb{S}:\nabla_x\mathbf{u} + \varrho\mathbf{f}\cdot\mathbf{u}\,\mathrm{d}\mathbf{x}\,\mathrm{d}t \qquad (4.47)$$

for any $\tau \geq 0$ (cf. Section 4.2). Clearly, relations (4.46), (4.47) are not compatible with (4.45).

In a weak formulation "with defect measure", one equation *(1.36) is replaced by two* inequalities *(4.41) and (4.45).*

4.3.3 Renormalized solutions

Consider a real function h such that

$$h \in C^2[0,\infty), h(0) = 1, h \text{ non-increasing on } [0,\infty), \lim_{z\to\infty} h(z) = 0,$$

$$h''(z)h(z) \geq 2(h'(z))^2 \quad \text{for all } z \geq 0. \qquad (4.48)$$

Multiplying the inequality (4.38) by $h(\vartheta)$ yields

$$\partial_t(\varrho Q_h(\vartheta)) + \mathrm{div}_x(\varrho Q_h(\vartheta)\mathbf{u}) - \Delta\mathcal{K}_h(\vartheta)$$

$$\geq h(\vartheta)\left(\mathbb{S}:\nabla_x\mathbf{u} - \vartheta p_\vartheta\mathrm{div}_x\mathbf{u}\right) - h'(\vartheta)\kappa(\vartheta)|\nabla_x\vartheta|^2,$$

where the functions Q_h, $\mathcal{K}_h$ are defined through

$$Q_h(\vartheta) \equiv \int_0^\vartheta Q'(z)h(z)\,\mathrm{d}z = \int_0^\vartheta c_v(z)h(z)\,\mathrm{d}z, \quad \mathcal{K}_h(\vartheta) = \int_0^\vartheta \kappa(z)h(z)\,\mathrm{d}z.$$

$$(4.49)$$

Note that

$$\mathrm{div}_x(\kappa(\vartheta)\nabla_x\vartheta)h(\vartheta) = \Delta\mathcal{K}_h(\vartheta) - \kappa(\vartheta)h'(\vartheta)|\nabla_x\vartheta|^2,$$

where, by virtue of (4.48), $h'(\vartheta) \leq 0$ for any $\vartheta \geq 0$.

Definition 4.4 We shall say that a function ϑ is a *renormalized solution* of (4.38), (4.39) supplemented with the initial condition

$$\vartheta(0) \geq \vartheta_0$$

if

$$\vartheta \geq 0, \quad \varrho Q(\vartheta) \in L^2(0, T; L^{q_2}(\Omega)), \quad q_2 > \frac{2N}{N+2};$$

$$\mathbf{u} \in L^2(0, T; W_0^{1,2}(\Omega; R^N));$$

$$\vartheta^{\alpha/2} \in L^2(0, T; W^{1,2}(\Omega)), \quad \mathcal{K}(\vartheta) \in L^1((0, T) \times \Omega);$$

$$\vartheta p_\vartheta \in L^2((0, T) \times \Omega), \quad \mathbb{S} \in L^2(0, T; L^2(\Omega; R^{N^2}));$$

and the integral inequality

$$\int_0^T \int_\Omega \varrho Q_h(\vartheta)\partial_t \varphi + \varrho Q_h(\vartheta)\mathbf{u} \cdot \nabla_x \varphi + \mathcal{K}_h(\vartheta)\Delta\varphi \, \mathrm{dx} \, \mathrm{dt}$$

$$\leq \int_0^T \int_\Omega h(\vartheta)\left(\vartheta p_\vartheta \mathrm{div}_x \mathbf{u} - \mathbb{S} : \nabla_x \mathbf{u}\right)\varphi \, \mathrm{dx} \, \mathrm{dt}$$

$$+ \int_0^T \int_\Omega h'(\vartheta)\kappa(\vartheta)|\nabla_x\vartheta|^2 \varphi \, \mathrm{dx} \, \mathrm{dt} - \int_\Omega \varrho(0+)Q_h(\vartheta_0)\varphi(0) \, \mathrm{dx} \quad (4.50)$$

holds for any h satisfying (4.48) and any test function φ satisfying (4.42). The functions Q_h, $\mathcal{K}_h$ are given by (4.49).

Note that the hypothesis $\varrho Q(\vartheta) \in L^2(0, T; L^{q_2}(\Omega))$ does not follow, in general, from the *a priori* estimates presented in Theorem 3.1. However, if we impose the additional growth condition (3.6) on Q, then

$$Q(\vartheta) \text{ is bounded in } L^2(0, T; W^{1,2}(\Omega)) \quad (4.51)$$

in accordance with (3.49). In particular, assuming $\varrho \in L^\infty(0, T; L^\gamma(\Omega))$ with $\gamma > \frac{N}{2}$ we have

$$\varrho Q(\vartheta) \in L^2(0, T; L^{q_2}(\Omega)), \quad \text{with } q_2 = m_2 > \frac{2N}{N+2},$$

where m_2 is the same as in (3.46).

Unfortunately, we are not able to establish any *weak stability* result for the renormalized solutions of (4.38)–(4.40). For the time being, we content ourselves with a straightforward consequence of Corollary 2.2 to illuminate the role of hypothesis (4.48).

Lemma 4.8   *Assume that*

$$\mathbf{u}_n \to \mathbf{u} \ \text{weakly in } L^2(0, T; W^{1,2}(\Omega; R^N)),$$

and

$$\mathbb{S}_n \to \mathbb{S} \ \text{weakly in } L^2(0, T; L^2(\Omega; R^{N^2})),$$

where $\mathbb{S}_n = \mathbb{S}_n[\nabla_x \mathbf{u}_n]$, $\mathbb{S} = \mathbb{S}[\nabla_x \mathbf{u}]$ *are determined by* (1.37) *with constant viscosity coefficients* μ *and* λ *satisfying* (1.38).

Let, moreover, $\{\vartheta_n\}_{n=1}^\infty$ *be a sequence of non-negative functions such that*

$$\vartheta_n \to \vartheta \ \text{weakly in } L^1((0, T) \times \Omega).$$

Then $h(\vartheta)\mathbb{S} : \nabla_x \mathbf{u}$ *is integrable on* $(0, T) \times \Omega$, *and*

$$\int_0^T \int_\Omega h(\vartheta)\mathbb{S} : \nabla_x \mathbf{u} \, \mathrm{d}\mathbf{x} \, \mathrm{d}t \leq \liminf_{n \to \infty} \int_0^T \int_\Omega h(\vartheta_n)\mathbb{S}_n : \nabla_x \mathbf{u}_n \, \mathrm{d}\mathbf{x} \, \mathrm{d}t$$

Proof   In accordance with hypothesis (1.38), the mapping

$$\nabla_x \mathbf{u} \mapsto \mathbb{S}[\nabla_x \mathbf{u}] : \nabla_x \mathbf{u} = \sum_{i,j} \frac{\mu}{2} \left(\frac{\partial u^i}{\partial x_j} + \frac{\partial u^j}{\partial x_i} \right)^2 + \lambda |\mathrm{div}_x \mathbf{u}|^2$$

is a positive quadratic form with respect to the partial derivatives $\partial_{x_j} u^i$, $i, j = 1, \dots, N$.

Consequently, by virtue of Corollary 2.2, it is enough to observe that a function

$$\Phi : (\vartheta, z) \mapsto \begin{cases} h(\vartheta)z^2 & \text{for } \vartheta \geq 0, \\[2mm] \infty & \text{if } \vartheta < 0 \end{cases}$$

is convex lower semi-continuous on R^2.

Computing the Hessian $\{\partial_{\vartheta,z}^2 \Phi\}$ we get

$$\left| \partial_{\vartheta,z}^2 \Phi \right| = 2z^2 \left(h''(\vartheta)h(\vartheta) - 2(h'(\vartheta))^2 \right) \geq 0$$

while

$$\mathrm{trace}\left[\partial_{\vartheta,z}^2 \Phi \right] = z^2 h''(\vartheta) + 2h(\vartheta) \geq 0$$

provided $\vartheta > 0$ and h satisfies (4.48). Thus the Hessian is positively definite for $\vartheta > 0$; therefore Φ is convex lower semi-continuous on R^2.

□

4.3.4 Time continuity of the thermal energy

In contrast to the density ϱ and the momentum $\varrho\mathbf{u}$, the instantaneous values of the thermal energy density $\varrho Q(\vartheta)$ are not continuous (not even weakly) with respect to time t.

Taking

$$\psi \in \mathcal{D}(0,T), \quad \psi \geq 0, \quad \text{and} \quad \eta \in C^{\infty}(\overline{\Omega}), \quad \eta \geq 0, \quad \nabla_x \eta \cdot \mathbf{n}|_{\partial\Omega} = 0,$$

and using $\varphi(t,\mathbf{x}) = \psi(t)\eta(\mathbf{x})$ as a test function in (4.41), we deduce that the integral averages of the thermal energy satisfy

$$t \mapsto \int_{\Omega} \varrho(t,\mathbf{x})Q(\vartheta)(t,\mathbf{x})\eta(\mathbf{x}) \, d\mathbf{x} = g_{\eta}^1(t) + g_{\eta}^2(t) \quad \text{for a.a. } t \in (0,T),$$

where g_{η}^1 is absolutely continuous and g_{η}^2 a non-decreasing function of time.

Accordingly, the right instantaneous value $(\varrho Q(\vartheta))(\tau+)$ and the left instantaneous value $(\varrho Q(\vartheta))(\tau-)$ determined through Definition 2.1 are non-negative measures satisfying

$$\varrho Q(\vartheta)(\tau-) \leq \varrho Q(\vartheta)(\tau+) \quad \text{for any } \tau \in (0,T).$$

In accordance with Definition 2.1, the instantaneous value $\varrho Q(\vartheta)(\tau)$ is the average of these two measures. Consequently, although the thermal energy density is not necessarily a continuous function of time, its points of discontinuity form at most countable subsets of the interval $[0,T]$.

As for the initial values, remember that any weak solution with a defect measure of the problem (4.38)–(4.40) is supposed to satisfy the energy inequality (4.45). On the other hand, if ϱ, $\varrho\mathbf{u}$ represent a variational solution of (1.34), (1.35) in the sense of Definitions 4.2, 4.3, we have

$$\varrho \in C([0,T]; L^1(\Omega)) \cap C([0,T]; L^{\gamma}_{\text{weak}}(\Omega)), \quad \varrho\mathbf{u} \in C([0,T]; L^{m_\infty}_{\text{weak}}(\Omega; R^N)).$$

Consequently, in accordance with our discussion in Section 4.2, the total *mechanical energy* E_{mech} defined through formula (4.36) is lower semi-continuous in t, in particular it follows from (4.45) that

$$\int_{\Omega} \chi_0 \, d\mathbf{x} = \int_{\Omega} \varrho(0)Q(\vartheta(0)) \, d\mathbf{x} \geq \text{ess} \lim_{t \to 0+} \sup \int_{\Omega} \varrho(t)Q(\vartheta)(t) \, d\mathbf{x}. \tag{4.52}$$

Indeed one has

$$E_{\text{mech}}(0) + \text{ess} \limsup_{t \to 0+} \int_{\Omega} \varrho(t)Q(\vartheta)(t) \, d\mathbf{x}$$

$$\leq \liminf_{t \to 0+} E_{\text{mech}}[\varrho(t), (\varrho\mathbf{u})(t)] + \text{ess} \limsup_{t \to 0+} \int_{\Omega} \varrho(t)Q(\vartheta)(t) \, d\mathbf{x}$$

$$\leq \text{ess} \limsup_{t \to 0+} E(t) \leq E(0) = E_{\text{mech}}(0) + \int_{\Omega} \chi_0 \, d\mathbf{x}.$$

Variational solutions

On the other hand, it follows from the integral inequality (4.41) that

$$\int_\Omega \chi_0 \eta \, d\mathbf{x} \le \int_\Omega \varrho(\tau)Q(\vartheta)(\tau)\eta \, d\mathbf{x} + \int_0^\tau h_\eta(t) \, dt,$$

$$\text{with } h_\eta \in L^1(0,T), \quad \text{for a.a. } \tau \in (0,T)$$

for any $\eta \in C^\infty(\overline{\Omega})$, $\eta \ge 0$, $\nabla_x \eta \cdot \mathbf{n}|_{\partial\Omega} = 0$; whence

$$\int_\Omega \chi_0 \eta \, d\mathbf{x} \le \operatorname*{ess\,lim\,inf}_{t\to 0+} \int_\Omega \varrho(t)Q(\vartheta(t))\eta \, d\mathbf{x}. \tag{4.53}$$

Combining (4.52) with (4.53) we infer that

$$\operatorname*{ess\,lim}_{t\to 0+} \varrho(t)Q(\vartheta)(t) = \chi_0 \text{ weakly-(*)} \quad \text{in } \mathcal{M}(\overline{\Omega}), \tag{4.54}$$

in particular the *instantaneous values* $\varrho Q(\vartheta)(t)$ in the sense of Definition 2.1 satisfy

$$\boxed{\varrho Q(\vartheta)(t) \to \chi_0 \text{ weakly-(*) in } \mathcal{M}(\overline{\Omega}) \quad \text{for } t \to 0+. \tag{4.55}}$$

Indeed, in accordance with (4.52), (4.53),

$$\int_\Omega \chi_0 \eta \, d\mathbf{x} = \int_\Omega \chi_0 \eta \, d\mathbf{x} + \int_\Omega \chi_0(1-\eta) \, d\mathbf{x} + \int_\Omega \chi_0(\eta-1) \, d\mathbf{x}$$

$$= \int_\Omega \chi_0 \, d\mathbf{x} + \int_\Omega \chi_0(\eta-1) \, d\mathbf{x}$$

$$\ge \operatorname*{ess\,lim\,sup}_{t\to 0+} \int_\Omega \varrho(t)Q(\vartheta(t)) \, d\mathbf{x}$$

$$+ \operatorname*{ess\,lim\,sup}_{t\to 0+} \int_\Omega \varrho(t)Q(\vartheta)(t)(\eta-1) \, d\mathbf{x}$$

$$\ge \operatorname*{ess\,lim\,sup}_{t\to 0+} \int_\Omega \varrho Q(\vartheta(t))\eta \, d\mathbf{x} \tag{4.56}$$

for any $0 \le \eta \le 1$. Relation (4.53), together with (4.56), yields (4.54).

4.3.5 Variational solutions

Definition 4.5 We shall say that a trio of functions ϱ, **u**, and ϑ is a *variational solution* of problem (1.36), (1.42) supplemented with the initial condition

$$\varrho Q(\vartheta)(0) = \chi_0 \quad \text{on } \Omega$$

if

(1) the density ϱ and the temperature ϑ are non-negative functions such that

$$\varrho Q(\vartheta) \in L^\infty(0,T;L^1(\Omega)) \cap L^2(0,T;L^{q_2}(\Omega)), \quad q_2 > \frac{2N}{N+2},$$

$$\log(\vartheta) \in L^2((0,T) \times \Omega).$$

(2) the velocity **u** and the stress tensor $\mathbb{S}$ determined by (1.37) satisfy

$$\mathbf{u} \in L^2(0,T;W_0^{1,2}(\Omega;R^N)), \quad \mathbb{S} \in L^2(0,T;L^2(\Omega;R^{N^2}));$$

(3) the thermal pressure ϑp_ϑ belongs to $L^2((0,T) \times \Omega)$, and $\mathcal{K}(\vartheta) \in L^1((0,T) \times \Omega)$;

(4) the functions ϱ, ϑ, and **u** satisfy the integral inequality (4.41) for any test function φ obeying (4.42);

(5) the *energy inequality* (4.45) holds for a.a. $\tau \in (0,T)$, with

$$E(0) = E_{\mathrm{mech}}[\varrho(0+),(\varrho u)(0+)] + \int_\Omega \chi_0 \, d\mathbf{x}$$

$$= \int_\Omega \frac{1}{2} \frac{|\varrho \mathbf{u}(0+)|^2}{\varrho(0+)} + \varrho(0+)P_e(\varrho(0+)) + \chi_0 \, d\mathbf{x}.$$

Remark The quantities $\varrho(0+)$, $(\varrho\mathbf{u})(0+)$ appearing in the formula for $E(0)$ are the right instantaneous values of the density ϱ and the momentum $\varrho\mathbf{u}$ at $t = 0$ introduced in Definition 2.1.

Unlike their counterparts appearing in Definition 4.2, variational solutions of the internal energy equation (1.36) need not be renormalized solutions in the sense of Definition 4.4. On the other hand, the concept of renormalized solutions will be used later in the construction of variational ones (see Chapters 6, 7).

4.4 Bibliographical notes

4.1 The renormalized solutions were introduced by DiPerna and P.-L. Lions [32] and further developed in the context of the Boltzmann equation (see [31]).

They are strongly related to the concept of entropy solutions introduced by Kruzhkov [73] for quasilinear conservation laws. The renormalized solutions have been successfully adapted by many authors in rather different areas of the theory of partial differential equation (see e.g. [11]). More recent applications to the Boltzmann equation can be found in the work of Alexandre and Villani [3], and [122].

Lemma 4.3 is probably due to Friedrichs; there are several variants available in the literature (see e.g. [84, 107]). Other results in this section may be found in [32].

It is interesting to examine the role of the "critical condition"

$$\varrho \in L^2(0, T; L^2(\Omega)),$$

which appears in both Proposition 4.1 and Corollary 4.1. Indeed admitting that the *a priori estimate*

$$\mathbf{u} \in L^2(0, T; W_0^{1,2}(\Omega; R^N))$$

is optimal, which seems to be a reasonable assumption given the amount of work done for compressible as well as incompressible fluids, we find Corollary 4.1 is the only tool for proving weak compactness of the family of all *renormalized solutions* of the continuity equation (1.34).

On the other hand, under the mere hypothesis $\gamma > N/2$, the square integrability of ϱ is an open problem. To remove this stumbling block, a new technique based on an oscillations defect measure will be developed in Chapter 6.

4.2   The validity of the energy equality seems to be intimately related to the question of uniqueness and regularity of weak solutions emanating from given initial data. This issue was discussed by Caffarelli et al. [17], Kozono and Sohr [71] among others in the context of the incompressible Navier–Stokes equations.

Several forms of energy inequality for the compressible barotropic case were discussed by P.-L. Lions in [85]. For a barotropic fluid, we have

$$\frac{\mathrm{d}}{\mathrm{d}t} E(t) + \int_\Omega \mathbb{S} : \nabla_x u \, \mathrm{d}\mathbf{x} \le \int_\Omega \varrho \mathbf{f} \cdot \mathbf{u} \, \mathrm{d}\mathbf{x}, \tag{4.57}$$

which can be understood as a "differential form" of (4.31). Of course, (4.57) is to be satisfied in the sense of distributions—in $\mathcal{D}'(0, T)$. The validity of (4.57) can be verified for rather general pressure–density barotropic state equations but under the condition that the underlying spatial domain has a compact Lipschitz boundary (see [50]). Boundary regularity is needed to construct an "inverse" to the div_x operator in order to improve the estimates on the pressure term. This issue will be discussed in the next section.

4.3   The variational formulation "with defect measure" as introduced for the thermal energy balance can be traced back to the work of DiPerna and P.-L.Lions on the Fokker–Planck–Boltzmann equation [30]. Here we follow the presentation of Alexandre and Villani [3].

Similar ideas in the context of the compressible Navier–Stokes equations were discussed by P.-L. Lions in Chapter 8 of [85]. The pressure is taken as the form $p = (\vartheta + \delta)\varrho^b$ with $\delta \geq 0$ and b "large enough", and (4.41) is replaced by the corresponding inequality for the entropy. Using this method, one loses control of several composed quantities, which are to be replaced by formal expressions in the resulting system of equations. These expressions coincide with the "correct" quantities provided the corresponding weak solution is smooth. Thus such a concept of weak solutions seems much closer to the so-called measure-valued solutions (see e.g. [89]) than to "classical" weak solutions considered in this book.

5

PRESSURE AND TEMPERATURE ESTIMATES

There are two main reasons why a weakly converging sequence $\{v_n\}_{n=1}^{\infty}$ in a Lebesgue space L^p may fail to be strongly convergent. The first difficulty is the possibility of very rapid fluctuations in the functions v_n. This is the problem of *oscillations*. Second, even if wild oscillations are excluded, we still cannot legitimately conclude strong convergence, as the mass (norm) in L^p may coalesce onto a set of zero Lebesgue measure. This is the problem of *concentration* we shall address in this chapter.

Up to now, the only available estimates on the *pressure p* have been deduced from the energy inequality (3.20) with the help of hypothesis (3.1), namely the elastic component p_e is bounded in $L^{\infty}(0,T; L^1(\Omega))$.

Since the *thermal pressure* ϑp_ϑ is bounded in $L^2((0,T) \times \Omega)$ in terms of the data (see (3.36)), we have

$$\|p\|_{L^1((0,T)\times\Omega)} \leq c(E_0, S_0, B_f, T). \tag{5.1}$$

The non-reflexive Banach space L^1 is not very convenient as bounded sequences are not necessarily weakly precompact; more precisely, concentration phenomena may occur to prevent bounded sequences in this space from converging weakly to an integrable function.

Similar problems arise when dealing with thermal energy inequality (4.41). Here, there are no bounds on the quantity $\mathcal{K}(\vartheta)$ whatsoever, not even in $L^1((0,T) \times \Omega)$. Indeed the only relevant estimate, namely (3.49) in Theorem 3.1, is not strong enough as, in accordance with hypothesis (3.3), $\mathcal{K}(\vartheta) \approx \vartheta^{\alpha+1}$ for large values of ϑ.

5.1 Local pressure estimates

The main way to get supplementary pressure estimates is very simple and natural, namely to "compute" the pressure p in the momentum equation (1.35) and use the energy estimates already available.

Applying the divergence operator to (1.35) we obtain

$$\Delta p = \mathrm{div}_x \mathrm{div}_x \mathbb{S} - \mathrm{div}_x \mathrm{div}_x (\varrho \mathbf{u} \otimes \mathbf{u}) + \mathrm{div}_x (\varrho \mathbf{f}) - \partial_t \mathrm{div}_x (\varrho \mathbf{u}) \tag{5.2}$$

Since we already know (see Theorem 3.1) that

$$
\left\{
\begin{array}{c}
\varrho \in L^\infty(0,T;L^\gamma(\Omega)), \\[2mm]
\varrho\mathbf{u} \in L^\infty(0,T;L^{m_\infty}(\Omega;R^N)), \\[2mm]
\varrho\mathbf{u}\otimes\mathbf{u} \in L^\infty(0,T;L^{c_2}(\Omega;R^{N^2})), \\[2mm]
\mathbb{S} \in L^2(0,T;L^2(\Omega;R^{N^2})),
\end{array}
\right\}
$$

the relation (5.2) can be viewed as an elliptic equation to be resolved with respect to the pressure p to obtain an estimate $p \in L^r((0,T)\times\Omega)$ for a certain $r > 1$. Obviously, there is a problem with the last term giving (formally) $\partial_t\Delta^{-1}\mathrm{div}_x\varrho\mathbf{u}$ for which there are no estimates available.

However, if ϱ is a renormalized solution of the continuity equation (1.34), we can use (5.2) to obtain

$$
\int_0^T\int_\Omega \psi\, pB(\varrho)\,\mathrm{d}\mathbf{x}\,\mathrm{d}t
$$

$$
\approx \text{bounded terms} + \int_0^T\int_\Omega \partial_t\psi(\Delta^{-1}\mathrm{div}_x)[\varrho\mathbf{u}]\,B(\varrho)\,\mathrm{d}\mathbf{x}\,\mathrm{d}t
$$

$$
+ \int_0^T\int_\Omega \psi\,(\Delta^{-1}\mathrm{div}_x)[\varrho\mathbf{u}]\,\partial_tB(\varrho)\,\mathrm{d}\mathbf{x}\,\mathrm{d}t,
$$

for any $\psi \in \mathcal{D}(0,T)$, where, by virtue of (4.13),

$$
\partial_tB(\varrho) = -b(\varrho)\mathrm{div}_x\mathbf{u} - \mathrm{div}_x(B(\varrho)\mathbf{u}), \quad \text{with } b \text{ given by (4.14).}
$$

In other words, if we succeed in making this formal procedure rigorous, we get pressure estimates of the form

$$
pB(\varrho) \text{ bounded in } L^1_{\mathrm{loc}}((0,T)\times\Omega)
$$

for a suitable function B.

To this end, let us introduce the *Riesz integral operator* $\mathcal{R}$:

$$
\mathcal{R}_i[v](\mathbf{x}) \equiv (-\Delta)^{-1/2}\partial_{x_i} = c\lim_{\varepsilon\to0}\int_{\{\varepsilon\leq|\mathbf{y}|\leq\frac{1}{\varepsilon}\}} v(\mathbf{x}-\mathbf{y})\frac{y_i}{|\mathbf{y}|^{N+1}}\,\mathrm{d}\mathbf{y}, \quad i=1,\ldots,N
$$

$$(5.3)$$

or, equivalently, in terms of its Fourier symbol,

$$
\mathcal{R}_i(\xi) = \frac{\mathrm{i}\xi_i}{|\xi|}, \quad \mathcal{R}_i[v] = \mathcal{F}^{-1}_{\xi\to x}\left[\frac{\mathrm{i}\xi_i}{|\xi|}\mathcal{F}_{x\to\xi}[v]\right], \quad i=1,\ldots,N,
$$

where $\mathcal{F}$ denotes the *Fourier transform*.

Now, the celebrated result of Calderón and Zygmund on singular integral operators may be formulated as follows:

Lemma 5.1   [18, Theorem 2]   *The Riesz operator $\mathcal{R}_i$, $i = 1, \ldots, N$ defined in (5.3) is a bounded linear operator on $L^p(R^N)$ for any $1 < p < \infty$.*

Next, we define another singular integral operator $\mathcal{A}$,

$$\mathcal{A}_i \equiv \partial_{x_i} \Delta^{-1}, \quad \text{or} \quad \mathcal{A}_i[v] = \mathcal{F}_{\xi \to x}^{-1} \left[\frac{-i\xi_i}{|\xi|^2} \mathcal{F}_{x \to \xi}[v] \right], \quad i = 1, \ldots, N, \qquad (5.4)$$

for which we claim the following result.

Lemma 5.2   (i) *Let $v \in L^1 \cap L^2(R^N)$, $N > 1$.*
Then $\mathcal{A}_i[v] \in (L^\infty \oplus L^2)(R^N)$, and

$$\| \mathcal{A}_i[v] \|_{(L^\infty \oplus L^2)(R^N)} \leq c\|v\|_{L^2 \cap L^1(R^N)}, \qquad (5.5)$$

$$\| \partial_{x_j} \mathcal{A}_i[v] \|_{L^p(R^N)} \leq c(p)\|v\|_{L^p(R^N)} \quad \textit{for any } 1 < p < \infty \qquad (5.6)$$

provided the right-hand side is finite.
    (ii) *The same conclusion holds if $N = 1$, and, in addition,*

$$\int_R v \, dx = 0, \quad \textit{and} \quad \text{supp}[v] \textit{ is contained in a ball of radius } r. \qquad (5.7)$$

If this is the case, the constant c in (5.5) depends on r.

Proof   Since $v \in L^1(R^N)$, the Fourier transform $\mathcal{F}_{x \to \xi}[v]$ is uniformly bounded, and, consequently,

$$\frac{\mathcal{F}_{x \to \xi}[v](\xi)}{|\xi|} \text{ is integrable in a neighborhood of } \xi = 0$$

provided $N > 1$. The same is true if $N = 1$ and (5.7) holds. Thus we have shown (5.5).
    As

$$\partial_{x_j} \mathcal{A}_i = -\mathcal{R}_j \mathcal{R}_i,$$

the estimate (5.6) follows directly from Lemma 5.1.

□

For η, $\xi \in \mathcal{D}(\Omega)$, $\Omega \subset R^N$, $N \geq 2$, $\psi \in \mathcal{D}(0,T)$, consider a test function

$$\varphi(t,\mathbf{x}) = \psi(t)\eta(\mathbf{x})\mathcal{A}[\xi(\cdot)\ B(t,\cdot)](t,\mathbf{x}),$$

where B is a bounded measurable function satisfying

$$\partial_t B + \mathrm{div}_x(B\mathbf{u}) = h \text{ in } \mathcal{D}'((0,T) \times \Omega) \quad \text{with } h \in L^2(0,T;W^{-1,q}(\Omega)). \quad (5.8)$$

Note that, by virtue of Lemmas 4.3, 5.1, we have

$$\mathrm{div}_x\varphi = \psi\nabla_x\eta \cdot \mathcal{A}[\xi B] + \psi\ \eta\xi\ B \in L^\infty((0,T) \times \Omega),$$
$$\partial_{x_j}\varphi_i = \psi\ \partial_{x_j}\eta\ \mathcal{A}_i[\xi B] + \psi\eta\ \partial_{x_j}\mathcal{A}_i[\xi B]$$
$$= \psi\ \partial_{x_j}\eta\ \mathcal{A}_i[\xi B] - \psi\ \eta\ \mathcal{R}_j\mathcal{R}_i[\xi B] \in L^\infty(0,T;L^p(\Omega)) \quad \text{for any } 1 < p < \infty,$$

and, by virtue of (5.8),

$$\partial_t\varphi = \partial_t\psi\ \eta\ \mathcal{A}[\xi B] + \psi\ \eta\ \mathcal{A}[\xi\partial_t B]$$
$$= \partial_t\psi\ \eta\ \mathcal{A}[\xi B] + \psi\ \eta\ \mathcal{A}[B\nabla_x\xi \cdot \mathbf{u}] - \psi\ \eta\ \mathcal{A}[\mathrm{div}_x(\xi B\mathbf{u})]$$
$$+ \psi\ \eta\ \mathcal{A}[\xi h] \in L^2(0,T;(L^{2^*} \oplus L^q)(\Omega, R^N)),$$

where 2^* is the critical Sobolev exponent,

$$2^* \text{ arbitrary finite for } N = 2, \quad 2^* = \frac{2N}{N-2} \text{ for } N \geq 3. \quad (5.9)$$

Under the hypotheses of Lemma 4.4, one can use φ as a test function for the momentum equation (4.27) to deduce the following result.

Lemma 5.3 *Let $\Omega \subset R^N$, $N \geq 2$ be an arbitrary domain, $\psi \in \mathcal{D}(0,T)$, η, $\xi \in \mathcal{D}(\Omega)$. Moreover, suppose that the quantities*

$$\varrho\mathbf{u} \in L^2(0,T;L^{m_2}(\Omega, R^N)), \quad \varrho\mathbf{u} \otimes \mathbf{u} \in L^2(0,T;L^{c_2}(\Omega; R^{N^2})),$$
$$p \in L^1((0,T) \times \Omega), \quad \mathbb{S} \in L^2(0,T;L^2(\Omega; R^{N^2}))$$

satisfy the momentum equation (1.35) in $\mathcal{D}'((0,T) \times \Omega)$. Finally, let B be a bounded measurable function satisfying (5.8) with

$$\mathbf{u} \in L^2(0,T;W^{1,2}(\Omega; R^N)), \quad h \in L^2(0,T;W^{-1,q}(\Omega)),$$

where

$$c_2 > 1, \quad \min\{2^*, q\} > (m_2)', \quad \frac{1}{m_2} + \frac{1}{m_2'} = 1, \quad \text{and } 2^* \text{ is given by (5.9).}$$

Then we have

$$\int_R \int_{R^N} \psi \eta \Big(\xi p B - \mathbb{S} : (\nabla_x \Delta^{-1} \nabla_x)[\xi B] \Big) \, dx \, dt = \sum_{j=1}^{7} I_j$$

$$+ \int_R \int_{R^N} \psi \mathbf{u} \Big[\xi B (\nabla_x \Delta^{-1} \nabla_x)[\eta \varrho \mathbf{u}] - (\nabla_x \Delta^{-1} \nabla_x)[\xi B] \eta \varrho \mathbf{u} \Big] \, dx \, dt \quad (5.10)$$

where

$$(\nabla_x \Delta^{-1} \nabla_x)_{i,j} \equiv -\mathcal{R}_i \mathcal{R}_j, \quad (\nabla_x \Delta^{-1} \mathrm{div}_x)_i \equiv -\mathcal{R}_i \mathcal{R}, \quad (5.11)$$

and

$$I_1 = \int_R \int_{R^N} \psi \, (\mathbb{S} \nabla_x \eta) \cdot \mathcal{A}[\xi B] \, dx \, dt,$$

$$I_2 = - \int_R \int_{R^N} \psi \, p \, \nabla_x \eta \cdot \mathcal{A}[\xi B] \, dx \, dt,$$

$$I_3 = - \int_R \int_{R^N} \psi \eta \, \varrho \mathbf{f} \cdot \mathcal{A}[\xi B] \, dx \, dt,$$

$$I_4 = - \int_R \int_{R^N} \psi \, ([\varrho \mathbf{u} \otimes \mathbf{u}] \nabla_x \eta) \cdot \mathcal{A}[\xi B] \, dx \, dt,$$

$$I_5 = - \int_R \int_{R^N} \psi \eta \, \varrho \mathbf{u} \cdot \mathcal{A}[B \nabla_x \xi \cdot \mathbf{u}] \, dx \, dt,$$

$$I_6 = - \int_R \int_{R^N} \partial_t \psi \, \eta \, \varrho \mathbf{u} \cdot \mathcal{A}[\xi B] \, dx \, dt,$$

$$I_7 = - \int_R \int_{R^N} \psi \eta \, \varrho \mathbf{u} \cdot \mathcal{A}[\xi h] \, dx \, dt.$$

If ϱ is a renormalized solution of the continuity equation (1.34), one can take the function $B = B(\varrho)$ and consider (4.13) instead of (5.8). This yields a new estimate implying, in particular, local boundedness of the pressure in a reflexive L^r-space with $r > 1$. Such an estimate plays a crucial role in the existence theory developed later in this book. Unlike the *a priori* estimates derived in Section 3 for regular (smooth) solutions, the validity of these *pressure estimates* extends to the class of distributional solutions.

Proposition 5.1 *Let $\Omega \subset R^N$, $N \geq 2$ be an arbitrary domain. Assume that ϱ, $\mathbf{u}$, p, and $\mathbb{S}$ solve the momentum equation (1.35) in $\mathcal{D}'((0,T) \times \Omega)$, and*

$$\varrho \in L^\infty(0,T;L^\gamma(\Omega)), \quad \varrho \geq 0, \quad \mathbf{u} \in L^2(0,T;W^{1,2}(\Omega;R^N)),$$

$$\mathbb{S} \in L^2(0,T;L^2(\Omega;R^{N^2})),$$

$$p \in L^1((0,T) \times \Omega), \quad p \text{ bounded from below},$$

where

$$\gamma > \frac{N}{2}. \tag{5.12}$$

Moreover, let ϱ be a renormalized solution of the continuity equation in the sense of Definition 4.1. Finally, assume that the kinetic energy is bounded, that is,

$$\sqrt{\varrho}|\mathbf{u}| \in L^\infty(0,T;L^2(\Omega)). \tag{5.13}$$

Then for any compact $O \subset ((0,T) \times \Omega)$, there is a constant $c = c(O)$ depending solely on the norm of the quantities ϱ, $\mathbf{u}$, and $\mathbb{S}$ in the aforementioned spaces such that

$$\int_O p\varrho^\omega \, d\mathbf{x} \, dt \leq c(O) \quad \text{for any } 0 < \omega < \min\left\{\frac{1}{N}, \frac{2}{N}\gamma - 1\right\}. \tag{5.14}$$

Proof In accordance with our hypotheses, Theorem 2.6, and Hölder's inequality (2.1), we have

$$\varrho\mathbf{u} \in L^2(0,T;L^{m_2}(\Omega;R^N)) \quad \text{with } m_2 = \frac{2N\gamma}{2N + \gamma(N-2)}, \tag{5.15}$$

$$\varrho\mathbf{u} \otimes \mathbf{u} \in L^2(0,T;L^{c_2}(\Omega;R^{N^2})) \quad \text{for } c_2 = \frac{2N\gamma}{N + 2\gamma(N-1)}, \tag{5.16}$$

where the exponents m_2, c_2 are the same as in Theorem 3.1. Obviously hypothesis (5.12) guarantees that

$$c_2 > 1 \quad \text{and} \quad 2^* > m_2' = \frac{2N\gamma}{N(\gamma-2) + 2\gamma}.$$

Thus one can use Lemma 5.3 for $B = B(T_k(\varrho))$ where B is a continuously differentiable function to be determined below and T_k are the cut-off functions defined in (2.11), (2.12). As this is a local result and ψ, η, and ξ have compact support, we may assume that Ω is a *bounded* domain with smooth boundary.

Accordingly, we have

$$\int_R \int_{R^N} \psi\eta\xi \; pB(T_k(\varrho)) \, \mathrm{d}\mathbf{x}\,\mathrm{d}t$$

$$= \int_R \int_{R^N} \psi\eta \; \mathbb{S} : (\nabla_x \Delta^{-1}\nabla_x)[\xi B(T_k(\varrho))] \, \mathrm{d}\mathbf{x}\,\mathrm{d}t + \sum_{j=1}^{7} I_j$$

$$+ \int_R \int_{R^N} \psi\mathbf{u}\Big[\xi B(T_k(\varrho))(\nabla_x \Delta^{-1}\nabla_x)[\eta\varrho\mathbf{u}]$$

$$- (\nabla_x \Delta^{-1}\nabla_x)[\xi B(T_k(\varrho))]\eta\varrho\mathbf{u}\Big] \, \mathrm{d}\mathbf{x}\,\mathrm{d}t,$$

where, by virtue of Lemma 5.1 and Hölder's inequality (2.1),

$$\left|\int_R \int_{R^N} \psi\eta \; \mathbb{S} : (\nabla_x \Delta^{-1}\nabla_x)[\xi B(T_k(\varrho))] \, \mathrm{d}\mathbf{x}\,\mathrm{d}t\right|$$

$$\leq c\|\mathbb{S}\|_{L^2(0,T;L^2(\Omega,R^{N^2}))}\|B(T_k(\varrho))\|_{L^2(0,T;L^2(\Omega))} \qquad (5.17)$$

provided

$$0 \leq \psi, \; \eta, \; \xi \leq 1.$$

Similarly, making use of (5.16) we arrive at

$$\left|\int_R \int_{R^N} \psi\mathbf{u} \cdot \left[(\nabla_x \Delta^{-1}\nabla_x)[\xi B(T_k(\varrho))]\, \eta\varrho\mathbf{u}\right.\right.$$

$$\left.\left. -\xi B(T_k(\varrho))\, (\nabla_x \Delta^{-1}\mathrm{div}_x)[\eta\varrho\mathbf{u}]\right] \, \mathrm{d}\mathbf{x}\,\mathrm{d}t\right|$$

$$\leq c\|\varrho\mathbf{u}\otimes\mathbf{u}\|_{L^2(0,T;L^{c_2}(\Omega;R^{N^2}))}\|B(T_k(\varrho))\|_{L^2(0,T;L^{c_2'}(\Omega))}, \qquad (5.18)$$

where

$$c_2' \equiv \frac{c_2}{c_2-1} = \frac{N\gamma}{2\gamma-N}.$$

Finally, with help of Lemma 5.2, the integrals $I_1,\ldots,I_7$ are estimated as follows:

$$|I_1| = \left|\int_R \int_{R^N} \psi \, (\mathbb{S}\nabla_x\eta) \cdot \mathcal{A}[\xi B(T_k(\varrho))] \, \mathrm{d}\mathbf{x}\,\mathrm{d}t\right|$$

$$\leq c(\nabla_x\eta) \, \|\mathbb{S}\|_{L^2(0,T;L^2(\Omega;R^{N^2}))} \, \|B(T_k(\varrho))\|_{L^2(0,T;L^2(\Omega))}; \qquad (5.19)$$

$$|I_2| = \left| \int_R \int_{R^N} \psi \, p \, \nabla_x \eta \cdot \mathcal{A}[\xi B(T_k(\varrho))] \, \mathrm{d}\mathbf{x} \, \mathrm{d}t \right|$$

$$\leq c(\nabla_x \eta) \, \|p\|_{L^1((0,T)\times\Omega)} \, \| \mathcal{A}[\xi B(T_k(\varrho))] \|_{L^\infty(0,T;L^\infty(\Omega))}$$

$$\leq c(\nabla_x \eta, \xi) \, \|p\|_{L^1((0,T)\times\Omega)} \, \|B(T_k(\varrho))\|_{L^\infty(0,T;L^q(\Omega))} \quad \text{for } q > N, \quad (5.20)$$

where we have used the imbedding $W^{1,q}(\Omega) \subset L^\infty(\Omega)$ (see Theorem 2.7).

Using the same argument we obtain

$$|I_3 + I_4 + I_6| \quad = \left| \int_R \int_{R^N} \psi \eta \, \varrho \mathbf{f} \cdot \mathcal{A}[\xi B] \, \mathrm{d}\mathbf{x} \, \mathrm{d}t \right.$$

$$+ \int_R \int_{R^N} \psi \, ([\varrho \mathbf{u} \otimes \mathbf{u}] \nabla_x \eta) \cdot \mathcal{A}[\xi B] \, \mathrm{d}\mathbf{x} \, \mathrm{d}t$$

$$+ \left. \int_R \int_{R^N} \partial_t \psi \, \eta \, \varrho \mathbf{u} \cdot \mathcal{A}[\xi B] \, \mathrm{d}\mathbf{x} \, \mathrm{d}t \right|$$

$$\leq c(\partial_t \psi) \Big(\|\mathbf{f}\|_{L^\infty(0,T;L^\infty(\Omega))} \|\varrho\|_{L^\infty(0,T;L^1(\Omega))}$$

$$+ \|\varrho \mathbf{u} \otimes \mathbf{u}\|_{L^2(0,T;L^1(\Omega;R^{N^2}))}$$

$$+ \|\varrho \mathbf{u}\|_{L^\infty(0,T;L^1(\Omega))} \Big) \|B(T_k(\varrho))\|_{L^\infty(0,T;L^q(\Omega))}, \quad q > N. \quad (5.21)$$

Next, by virtue of (5.15), Lemma 5.2, and Hölder's inequality (2.1),

$$|I_5| = \left| \int_R \int_{R^N} \psi \eta \, \varrho \mathbf{u} \cdot \mathcal{A}[B \nabla_x \xi \cdot \mathbf{u}] \, \mathrm{d}\mathbf{x} \, \mathrm{d}t \right|$$

$$\leq c(\nabla_x \xi) \|\varrho \mathbf{u}\|_{L^2(0,T;L^{m_2}(\Omega;R^N))} \|B(T_k(\varrho)) \mathbf{u}\|_{L^2(0,T;L^{m_2'}(\Omega;R^N))}$$

$$\leq c(\xi) \|\varrho \mathbf{u}\|_{L^2(0,T;L^{m_2}(\Omega;R^N))} \|\mathbf{u}\|_{L^2(0,T;W^{1,2}(\Omega;R^N))}$$

$$\|B(T_k(\varrho))\|_{L^\infty(0,T;L^{c_2'}(\Omega))}. \quad (5.22)$$

Finally, we have

$$|I_7| = \left| \int_R \int_{R^N} \psi \eta \, \varrho \mathbf{u} \cdot \mathcal{A}[\xi h] \, \mathrm{d}\mathbf{x} \, \mathrm{d}t \right|,$$

with

$$h = \Big(B(T_k(\varrho)) - B'(T_k(\varrho)) T_k'(\varrho) \varrho \Big) \mathrm{div}_x \mathbf{u}.$$

Consequently, we can use Lemma 5.2 together with Hölder's inequality (2.1) and Theorem 2.6 to obtain

$$|I_7| \leq c \|\varrho \mathbf{u}\|_{L^2(0,T;L^{m_2}(\Omega;R^N))} \|\mathcal{A}[\xi h]\|_{L^2(0,T;L^{m_2'}(\Omega))}$$

$$\leq c \|\varrho \mathbf{u}\|_{L^2(0,T;L^{m_2}(\Omega;R^N))} \|h\|_{L^2(0,T;L^q(\Omega))}, \quad q = \frac{2N\gamma}{N\gamma - 2N + 4\gamma}, \quad (5.23)$$

where

$$\|h\|_{L^2(0,T;L^q(\Omega))}$$
$$\leq \|\mathrm{div}_x\mathbf{u}\|_{L^2(0,T;L^2(\Omega))}\|B(T_k(\varrho)) - B'(T_k(\varrho))T_k'(\varrho)\varrho\|_{L^\infty(0,T;L^{c_2'}(\Omega))}.$$

To conclude the proof, we take B a smooth function such that

$$B(z) = z^\omega \quad \text{for } z \geq 1, \quad 0 < \omega < \min\left\{\frac{1}{N}, \frac{2}{N}\gamma - 1\right\},$$

let $k \to \infty$ in (5.17)–(5.23), and use the Lebesgue theorem to arrive at (5.14). Note that for the integrals in (5.17), (5.19) to be bounded we need $\omega \leq \gamma/2$; (5.20), (5.21) require $\omega < \gamma/N$ while we have to use $\omega \leq (2/N)\gamma - 1$ in (5.19), (5.22) and (5.23).

□

Note that there is a qualitative difference between the *a priori* estimates discussed in Section 3 and the *pressure estimates* obtained in Proposition 5.1. *A priori* estimates are formal, derived under the hypothesis that the solutions in question are classical and satisfy the equations pointwise while the pressure estimate (5.14) holds for *any* weak solution belonging to a suitable class (determined by *a priori* estimates).

The relation (5.14) is very important because it prevents formation of concentration points and makes it possible to show weak sequential stability (compactness) for the momentum equation (1.35).

5.2 Temperature estimates

We have already shown *a priori* estimates on the temperature ϑ in the space $L^2(0,T;W^{1,2}(\Omega))$ resulting from entropy equation (1.26) on condition that the heat conductivity coefficient κ grows at least as a quadratic function of ϑ. More specifically, if κ satisfies hypothesis (3.3), we obtained

$$\vartheta^{\alpha+1-\omega/2} \text{ bounded in } L^2(I;W^{1,2}(\Omega)) \tag{5.24}$$

for any $0 < \omega < 1$ on condition that the heat flux $\mathbf{q}$ satisfies the homogeneous boundary conditions (1.42) (see Theorem 3.1).

The relation (5.24) combined with the imbedding $W^{1,2} \subset L^{2^*}(\Omega)$ imply

$$\left.\begin{array}{l} \vartheta^{\alpha+1-\omega} \text{ bounded in } L^1(0,T;L^{N/N-2}(\Omega)) \quad \text{if } N \geq 3, \\[2ex] \vartheta^{\alpha+1-\omega} \text{ bounded in } L^1(0,T;L^p(\Omega)) \text{ for any } 1 \leq p < \infty \quad \text{if } N = 2. \end{array}\right\} \tag{5.25}$$

By virtue of the energy estimates, we have the function $\varrho\vartheta$ bounded in $L^\infty(0,T;L^1(\Omega))$, which together with (5.25) and Hölder's inequality, yields

$$\|\vartheta^{1+\alpha+1/N}(t)\|_{L^1(K)} = \|\vartheta^{1+\alpha-1/N}\|_{L^{N/N-2}(K)}\|\vartheta^{2/N}\|_{L^{N/2}(K)}$$

$$\leq \|\vartheta^{1+\alpha-1/N}\|_{L^{N/N-2}(K)}\|\vartheta\|_{L^1(K)}^{2/N} \quad \text{for any } K \subset \Omega;$$

in particular,

$$\vartheta^{1+\alpha} \text{ is bounded in } L^p\Big(\{\varrho \geq \varepsilon\}\Big) \quad \text{for a certain } p > 1 \tag{5.26}$$

provided $\varepsilon > 0$ and $N \geq 3$. A similar argument yields (5.26) also for $N = 2$.
Now, we have

$$\int_{\{\varrho(t) \geq \varepsilon\}} \varrho(t)\,\mathrm{d}\mathbf{x} \geq M - \varepsilon|\Omega|,$$

where $M = \int_\Omega \varrho(t)\,\mathrm{d}\mathbf{x}$ is the total mass—a constant of motion.
On the other hand, by virtue of Hölder's inequality (2.1),

$$\int_{\{\varrho(t) \geq \varepsilon\}} \varrho(t)\,\mathrm{d}\mathbf{x} \leq |\{\varrho \geq \varepsilon\}|^{(\gamma-1)/\gamma}\|\varrho(t)\|_{L^\gamma(\Omega)},$$

where, in accordance with Theorem 3.1, we can assume

$$\operatorname*{ess\,sup}_{t \in [0,T]} \|\varrho(t)\|_{L^\gamma(\Omega)} \leq c.$$

Consequently, there exists a non-negative function $\omega = \omega(\varepsilon)$, such that

$$|\{\varrho \geq \varepsilon\}| \geq \omega(\varepsilon) > 0 \text{ independently of } t \in [0,T] \text{ provided } 0 \leq \varepsilon < \frac{M}{|\Omega|}. \tag{5.27}$$

As the next step, fix $0 < \varepsilon < M/2|\Omega|$ and take a function $B \in C^\infty(R)$ such that

$$B : R \to R \text{ non-increasing,} \quad B(z) = 0 \text{ for } z \leq \varepsilon, \quad B(z) = -1 \text{ for } z \geq 2\varepsilon.$$

For each $t \in (0,T)$, one can solve the Neumann problem

$$\Delta\eta(t) = B(\varrho(t)) - \frac{1}{|\Omega|}\int_\Omega B(\varrho(t))\,\mathrm{d}\mathbf{x} \text{ in } \Omega, \quad \nabla\eta \cdot \mathbf{n} = \text{ on } \partial\Omega, \quad \int_\Omega \eta(t)\,\mathrm{d}\mathbf{x} = 0. \tag{5.28}$$

Now, the idea is to use

$$\varphi(t,\mathbf{x}) = \psi(t)[\eta(t,\mathbf{x}) - \underline{\eta}], \quad \text{with } \underline{\eta} = \inf_{t \in (0,T), \mathbf{x} \in \Omega} \eta(t,\mathbf{x}),$$

where $\psi \in \mathcal{D}(0,T)$,

$$0 \leq \psi \leq 1, \ \psi \text{ non-decreasing on } (0,a], \ \psi \text{ non-increasing on } [a,T),$$

as a test function in (4.41). Assuming ϱ is a renormalized solution of (1.34) on $(0, T) \times R^N$, the time derivative of η satisfies

$$\partial_t(\Delta\eta) = \partial_t B(\varrho) - \frac{1}{|\Omega|} \int_\Omega \partial_t B(\varrho)\,\mathrm{d}x$$

$$= -\mathrm{div}_x(B(\varrho)\mathbf{u}) - b(\varrho)\mathrm{div}_x\mathbf{u} + \frac{1}{|\Omega|} \int_\Omega b(\varrho)\mathrm{div}_x\mathbf{u}\,\mathrm{d}x \quad \text{in } \mathcal{D}'((0, T) \times \Omega),$$

where

$$b(\varrho) = B'(\varrho)\varrho - B(\varrho) \text{ is uniformly bounded.}$$

More precisely, there is $\mathbf{g} \in [L^2((0, T) \times \Omega)]^N$ such that the integral identity

$$\int_0^T \int_\Omega \partial_t\psi\,\nabla_x\eta \cdot \nabla_x\varphi\,\mathrm{d}x\,\mathrm{d}t = \int_0^T \int_\Omega \psi\mathbf{g} \cdot \nabla_x\varphi\,\mathrm{d}x\,\mathrm{d}t$$

holds for any $\psi \in \mathcal{D}(0, T)$ and $\varphi \in W^{1,2}(\Omega)$. Thus we get

$$\|\eta(t_2) - \eta(t_1)\|_{W^{1,2}(\Omega)} \leq \int_{t_1}^{t_2} \|\mathbf{g}\|_{L^2(\Omega; R^N)}\,\mathrm{d}t \quad \text{for a.a. } t_1, t_2 \in [0, T].$$

Now, we can use the following result.

Theorem 5.1 [19, Theorem 1.4.40] *Let X be a reflexive Banach space and $1 \leq p < \infty$. Let $v \in L^p(0, T; X)$.*

Then $\partial_t v$ belongs to the space $L^p(0, T; X)$ if and only if there exists a function $g \in L^p(0, T)$ such that

$$\|v(t_1) - v(t_2)\|_X \leq \left| \int_{t_1}^{t_2} g(t)\,\mathrm{d}t \right| \quad \text{for a.a. } t_1, t_2 \in [0, T].$$

In addition, we have

$$\|\partial_t v\|_{L^p(0,T;X)} \leq \|g\|_{L^p(0,T)}.$$

Applying Theorem 5.1, and this is by no means optimal, we have

$$\partial_t\eta \text{ in } L^2(0, T; W^{1,2}(\Omega)). \tag{5.29}$$

Thus φ satisfies the hypotheses of Lemma 4.7 and can be taken as a test function in (4.41) to obtain:

$$\int_0^T \int_\Omega \psi\mathcal{K}(\vartheta)\left(B(\varrho) - \frac{1}{|\Omega|} \int_\Omega B(\varrho)\,\mathrm{d}x\right)\mathrm{d}x\,\mathrm{d}t \leq \sum_{j=1}^5 I_j, \tag{5.30}$$

with integral terms

$$I_1 \equiv \int_0^T \int_\Omega \partial_t \psi (\underline{\eta} - \eta) \, \varrho Q(\vartheta) \, \mathrm{d}\mathbf{x} \, \mathrm{d}t,$$

$$I_2 \equiv - \int_0^T \int_\Omega \psi \, \partial_t \eta \, \varrho Q(\vartheta) \, \mathrm{d}\mathbf{x} \, \mathrm{d}t,$$

$$I_3 \equiv - \int_0^T \int_\Omega \psi \, \varrho Q(\vartheta) \mathbf{u} \cdot \nabla_x \eta \, \mathrm{d}\mathbf{x} \, \mathrm{d}t,$$

$$I_4 \equiv \int_0^T \int_\Omega \psi (\underline{\eta} - \eta) \, \mathbb{S} : \nabla_x \mathbf{u} \, \mathrm{d}\mathbf{x} \, \mathrm{d}t,$$

$$I_5 \equiv \int_0^T \int_\Omega \psi (\eta - \underline{\eta}) \, \vartheta \, p_\vartheta \mathrm{div}_x \mathbf{u} \, \mathrm{d}\mathbf{x} \, \mathrm{d}t.$$

The leading term on the left-hand side can be written as

$$\int_0^T \int_\Omega \psi \mathcal{K}(\vartheta) \Big(B(\varrho) - \frac{1}{|\Omega|} \int_\Omega B(\varrho) \, \mathrm{d}\mathbf{x} \Big) \, \mathrm{d}\mathbf{x} \, \mathrm{d}t$$

$$= \int_{\{\varrho < \varepsilon\}} \psi \mathcal{K}(\vartheta) \Big(B(\varrho) - \frac{1}{|\Omega|} \int_\Omega B(\varrho) \, \mathrm{d}\mathbf{x} \Big) \mathrm{d}\mathbf{x} \, \mathrm{d}t$$

$$+ \int_{\{\varrho \geq \varepsilon\}} \psi \mathcal{K}(\vartheta) \Big(B(\varrho) - \frac{1}{|\Omega|} \int_\Omega B(\varrho) \, \mathrm{d}\mathbf{x} \Big) \mathrm{d}\mathbf{x} \, \mathrm{d}t, \qquad (5.31)$$

where the second integral on the right-hand side is bounded in view of (5.26).
    On the other hand,

$$- \frac{1}{|\Omega|} \int_\Omega B(\varrho) \, \mathrm{d}\mathbf{x} \geq - \frac{1}{|\Omega|} \int_{\{\varrho \geq 2\varepsilon\}} B(\varrho) \, \mathrm{d}\mathbf{x} = \frac{|\{\varrho \geq 2\varepsilon\}|}{|\Omega|} \geq \frac{\omega(2\varepsilon)}{|\Omega|} > 0,$$

where we have used (5.27). Consequently,

$$\int_{\{\varrho < \varepsilon\}} \psi \mathcal{K}(\vartheta) \Big(B(\varrho) - \frac{1}{|\Omega|} \int_\Omega B(\varrho) \, \mathrm{d}\mathbf{x} \Big) \mathrm{d}\mathbf{x} \, \mathrm{d}t$$

$$\geq \frac{\omega(2\varepsilon)}{|\Omega|} \int_{\{\varrho < \varepsilon\}} \psi \mathcal{K}(\vartheta) \mathrm{d}\mathbf{x} \, \mathrm{d}t.$$

    The relations (5.26), (5.30) will yield estimates on the temperature ϑ in the space $L^{\alpha+1}((0,T) \times \Omega)$ provided we show that the integrals $I_j, j = 1, \ldots, 5$ are bounded.

To this end, we utilize Hölder's inequality (2.1) to obtain

$$|I_1| = \left| \int_0^T \int_\Omega \partial_t \psi (\underline{\eta} - \eta) \varrho Q(\vartheta) \, \mathrm{d}\mathbf{x} \, \mathrm{d}t \right|$$

$$\leq 2\|\eta\|_{L^\infty((0,T)\times\Omega)} \|\varrho Q(\vartheta)\|_{L^\infty(0,T;L^1(\Omega))};$$

$$|I_2| = \left| \int_0^T \int_\Omega \psi \partial_t \eta \, \varrho Q(\vartheta) \, \mathrm{d}\mathbf{x} \, \mathrm{d}t \right|$$

$$\leq \|\varrho\|_{L^\infty(0,T;L^\gamma(\Omega))} \|Q(\vartheta)\|_{L^2(0,T;L^{2^*}(\Omega))} \|\partial_t \eta\|_{L^2(0,T;L^{2^*}(\Omega))}$$

provided $\gamma > N/2$;

$$|I_3| = \left| \int_0^T \int_\Omega \psi \, \varrho Q(\vartheta) \mathbf{u} \cdot \nabla_x \eta \, \mathrm{d}\mathbf{x} \, \mathrm{d}t \right|$$

$$\leq \|\varrho\|_{L^\infty(0,T;L^\gamma(\Omega))} \|Q(\vartheta)\|_{L^2(0,T;L^{2^*}(\Omega))}$$

$$\times \|\mathbf{u}\|_{L^2(0,T;L^{2^*}(\Omega;R^N))} \|\nabla_x \eta\|_{L^\infty((0,T)\times\Omega)};$$

$$|I_4| = \left| \int_0^T \int_\Omega \psi(\underline{\eta} - \eta) \, \mathbb{S} : \nabla_x \mathbf{u} \, \mathrm{d}\mathbf{x} \, \mathrm{d}t \right|$$

$$\leq 2\|\mathbf{u}\|^2_{L^2(0,T;W^{1,2}(\Omega;R^N))} \|\eta\|_{L^\infty((0,T)\times\Omega)};$$

and

$$|I_5| = \left| \int_0^T \int_\Omega \psi(\eta - \underline{\eta}) \, \vartheta p_\vartheta \mathrm{div}_x \mathbf{u} \, \mathrm{d}\mathbf{x} \, \mathrm{d}t \right|$$

$$\leq \|\vartheta p_\vartheta\|_{L^2(0,T;L^2(\Omega))} \|\mathbf{u}\|_{L^2(0,T;W^{1,2}(\Omega;R^N))} \|\eta\|_{L^\infty((0,T)\times\Omega)}.$$

As η solves the elliptic equation (5.28) with a bounded right-hand side, uniform boundedness of $\nabla_x \eta$ (and that of η) follows from standard L^p-estimates (see, e.g., [2]) provided $\partial\Omega$ is sufficiently regular. Moreover, $\partial_t \eta$ is estimated by (5.29), and $Q(\vartheta)$, ϑp_ϑ satisfy (4.51) and (3.36) for any variational solution of the problem.

Consequently, we have shown that

$$\vartheta \text{ is bounded in } L^{\alpha+1}((0,T) \times \Omega), \tag{5.32}$$

which is equivalent to integrability of $\mathcal{K}(\vartheta)$.

We do not know how to improve (5.32) to get boundedness of $\mathcal{K}(\vartheta)$ in a reflexive space like L^r with $r > 1$. In the existence proof, however, we shall work with the renormalized inequality (4.50), where (5.32) is sufficient to make the compositions $\mathcal{K}_h(\vartheta)$ equi-integrable and thus weakly L^1-compact (cf. Theorem 2.10).

5.3 Bibliographical notes

5.1 The *local pressure estimates* were obtained by P.-L. Lions [85]. Formulas similar to (5.10) appear in [83].

It is interesting to note that these estimates hold, in fact, for the whole domain Ω provided its boundary is Lipschitz (see [52]). The main way to show this is to replace the operator $\mathcal{A}_i = \partial_{x_i}\Delta^{-1}$ used in Section 5.1 by a general operator $\mathcal{B}$ which corresponds in a certain sense to the inverse of div_x. More specifically, consider the problem

$$\mathrm{div}_x \, \mathbf{v} = g - \frac{1}{|\Omega|}\int_\Omega g\,\mathrm{d}\mathbf{x} \text{ on } \Omega, \quad \mathbf{v}|_{\partial\Omega} = 0. \tag{5.33}$$

The equation (5.33) has been studied by many authors. In particular, it can be shown (see [12]) that (5.33) admits a solution operator $\mathcal{B} : g \mapsto \mathbf{v}$ enjoying the following properties:

- $\mathcal{B}$ is a bounded linear operator from $L^p(\Omega)$ into $W_0^{1,p}(\Omega)$ for any $1 < p < \infty$;
- the function $\mathbf{v} = \mathcal{B}[g]$ solves the problem (5.33);
- if a function $g \in L^p(\Omega)$ can be written in the form $g = \mathrm{div}_x\mathbf{h}$ where $\mathbf{h} \in L^r(\Omega; R^N)$, $\mathbf{h}\cdot\mathbf{n} = 0$ on $\partial\Omega$, then

$$\|\mathcal{B}[g]\|_{L^r(\Omega)} \le c(p,r)\|\mathbf{h}\|_{L^r(\Omega)}.$$

The proof of the existence of $\mathcal{B}$ as well as the above properties can be found in [58] or [13]. Using $\mathcal{B}$ in place of $\mathcal{A}$ one can show that the local pressure estimates established in Proposition 5.1 hold for the whole domain Ω provided $\partial\Omega$ is Lipschitz (see [52]). An alternative method to obtain the same result was proposed in [86].

Validity of the pressure estimates "up to the boundary" is essential for the existence of weak solutions as presented in [85] (see [86]). It is worth noting that these estimates yield

$$\varrho \in L^{\gamma+\omega}((0,T)\times\Omega), \quad \text{with } \omega = \frac{2}{N}\gamma - 1. \tag{5.34}$$

On the other hand, the mathematical theory developed in [85] yields global existence results for the barotropic case under the additional hypotheses

$$\gamma \ge \tfrac{3}{2} \text{ for } N = 2, \quad \gamma \ge \tfrac{9}{5} \text{ if } N = 3.$$

Substituting these values of γ in (5.34) we recover once more the "critical condition"

$$\varrho \in L^2(0,T;L^2(\Omega)).$$

The existence theory developed in this book requires only local pressure estimates with $\omega > 0$ as stated in Proposition 5.1, which makes it applicable to problems in optimal shape design, where smoothness of the boundary is not known *a priori* (cf. [49]), or the problem of rigid bodies moving in the fluid, where the boundary is not smooth at all (see [48]).

For barotropic flows with $p \approx a\varrho^\gamma$, the estimates obtained with the help of the operator $\mathcal{B}$ can be used to show the existence of *bounded absorbing sets*. We report the following result (see Theorem 3.5 in [46]):

Theorem 5.2 *Let $\Omega \subset R^N$ be a bounded domain with Lipschitz boundary, $N = 2, 3$. Assume the pressure $p = p(\varrho)$ satisfies the isentropic state equation*

$$p(\varrho) = a\varrho^\gamma, \quad a > 0,$$

$$\gamma > 1 \text{ for } N = 2, \quad \gamma > \tfrac{5}{3} \text{ for } N = 3.$$

Let $\mathbf{f}$ be a bounded measurable function on $(0, \infty) \times \Omega$.

Then there exists a constant E_∞ depending solely on the (given) total mass M having the following property:

Given E_0, there exists a time $\tau = \tau(E_0)$ such that

$$E_{\text{mech}}[\varrho, \mathbf{u}](t) \le E_\infty \quad \text{for all } t > \tau$$

provided

$$\limsup_{t \to 0+} E_{\text{mech}}(t) \le E_0, \quad \int_\Omega \varrho \, d\mathbf{x} = M$$

and ϱ, $\mathbf{u}$ is a variational solution of the problem (1.34), (1.35), (1.41) on $(0, \infty) \times \Omega$ in the sense of Definitions 4.2, 4.3. Here, $E_{\text{mech}}[\varrho, \mathbf{u}]$ denotes the total mechanical energy introduced in (4.35).

Theorem 5.2 can be used to study the long-term behavior of solutions. Related results can be found in [51, 102, 103].

6

FUNDAMENTAL IDEAS

One of the main issues of the mathematical theory to be discussed in this book is the *propagation of density oscillations* The time evolution of the density ϱ is governed by the continuity equation (1.34), which is hyperbolic and linear with respect to ϱ. Therefore it is plausible to expect that oscillations in the initial data will be transported by the flow. Assuming weak convergence of initial data, specifically,

$$\varrho_n(0, \cdot) \to \varrho(0, \cdot) \text{ weakly in } L^1(\Omega)$$

one expects to recover only

$$\varrho_n(t, \cdot) \to \varrho(t, \cdot) \text{ weakly in } L^1(\Omega)$$

provided the corresponding velocity fields $\mathbf{u}_n$ are sufficiently regular. It is interesting to note that one of the reasons for this propagation of oscillations phenomenon to occur is, in fact, higher regularity of the velocity field due to the presence of viscosity. In the inviscid case, the oscillations in the density and velocity fields may mutually cancel, which yields a surprising regularizing effect at least when $N = 1$ as shown by DiPerna [28] (see also [87]).

As has been pointed out many times in this book, nonlinear compositions do not commute with weak limits. The failure of a sequence $\{\varrho_n\}_{n=1}^{\infty}$ to converge strongly in L^1 is often characterized by a *defect measure* of the form

$$\overline{B(\varrho)} - B(\varrho), \tag{6.1}$$

where B is a suitable convex function, ϱ is a weak limit of $\{\varrho_n\}_{n=1}^{\infty}$, and $\overline{B(\varrho)}$ stands for a weak limit of $\{B(\varrho_n)\}_{n=1}^{\infty}$. Indeed if B is strictly convex, we can use Theorem 2.11 to conclude that the quantity in (6.1) is always non-negative and vanishes precisely on the set where ϱ_n tends to ϱ pointwise, that is, where strong convergence takes place.

The defect measure introduced in (6.1) is particularly convenient when dealing with the *renormalized solutions* of the continuity equation; that means the functions ϱ_n, $\mathbf{u}_n$ together with their weak limits ϱ, $\mathbf{u}$ satisfy (4.13). If this is the case, we have

$$\partial_t\big(\overline{B(\varrho)} - B(\varrho)\big) + \text{div}_x\big(\big(\overline{B(\varrho)} - B(\varrho)\big)\mathbf{u}\big)$$
$$= b(\varrho)\text{div}_x\,\mathbf{u} - \overline{b(\varrho)\,\text{div}_x\,\mathbf{u}}, \tag{6.2}$$

where B, b are as in (4.14), and, in accordance with our convention, the symbol $b(\varrho) \operatorname{div}_x \mathbf{u}$ stands for a weak limit of $\{b(\varrho_n) \operatorname{div}_x \mathbf{u}_n\}_{n=1}^{\infty}$. Note that, by virtue of Corollary 2.1, we can assume

$$B(\varrho_n) \to \overline{B(\varrho)} \quad \text{in } C([0,T]; L_{\mathrm{weak}}^{\gamma}(\Omega))$$

provided B satisfies (4.15) and ϱ_n is bounded in $L^{\infty}(0,T; L^{\gamma}(\Omega))$. If, moreover,

$$\mathbf{u}_n \to \mathbf{u} \text{ weakly in } L^2(0,T; W^{1,2}(\Omega; R^N)),$$

we have

$$\overline{B(\varrho)}\, \mathbf{u} = \overline{B(\varrho)\, \mathbf{u}}$$

on condition that $\gamma > 2N/(N+2)$ since the Lebesgue space $L^{\gamma}(\Omega)$ is compactly imbedded into $W^{-1,2}(\Omega)$.

Consequently, in order to estimate the difference $\overline{B(\varrho)} - B(\varrho)$, we need a piece of information concerning the right-hand side of (6.2). However, this is a difficult task as both $\{b(\varrho_n)\}_{n=1}^{\infty}$ and $\{\operatorname{div}_x \mathbf{u}_n\}_{n=1}^{\infty}$ converge only weakly in, say, $L^2((0,T) \times \Omega)$. To cope with this apparent *lack of compactness*, we have to use the fact that ϱ_n, $\mathbf{u}_n$ satisfy also the equations of motion (1.35). This idea will be developed in what follows.

Remark   It seems interesting to observe that pointwise convergence together with uniform boundedness of the sequence $\{\operatorname{div}_x \mathbf{u}_n\}_{n=1}^{\infty}$ entails strong convergence of $\{\varrho_n\}_{n=1}^{\infty}$. More precisely, if $\operatorname{div}_x \mathbf{u}_n$ are bounded uniformly in $L^{\infty}((0,T) \times \Omega)$, and

$$\operatorname{div}_x \mathbf{u}_n \to \operatorname{div}_x \mathbf{u} \text{ a.a.} \quad \text{on } (0,T) \times \Omega,$$

we have

$$\overline{b(\varrho)\operatorname{div}_x \mathbf{u}} = \overline{b(\varrho)}\operatorname{div}_x \mathbf{u};$$

whence, by virtue of (6.2),

$$\frac{\mathrm{d}}{\mathrm{d}t} \int_{\Omega} \overline{\varrho^{\alpha}} - \varrho^{\alpha} \, \mathrm{d}\mathbf{x} \leq c \int_{\Omega} \overline{\varrho^{\alpha}} - \varrho^{\alpha} \, \mathrm{d}\mathbf{x} \quad \text{in } \mathcal{D}'(0,T)$$

for any $1 < \alpha < \gamma$. Thus one can use Gronwall's lemma (see Lemma 2.1) to conclude

$$\int_{\Omega} \overline{\varrho^{\alpha}}(t) - \varrho^{\alpha}(t) \, \mathrm{d}\mathbf{x} \leq \exp(Ct) \int_{\Omega} \overline{\varrho^{\alpha}}(0) - \varrho^{\alpha}(0) \, \mathrm{d}\mathbf{x}.$$

In other words, oscillations in $\{\varrho_n\}_{n=1}^{\infty}$ do not occur at any $t > 0$ unless they were present for $t = 0$.

6.1 The effective viscous pressure

The *effective viscous pressure* P_{eff} is strongly related to the expression appearing on the left-hand side of the identity (5.10) derived in Lemma 5.3. More precisely,

$$P_{\text{eff}} \equiv p - (\nabla_x \Delta^{-1} \nabla_x) : \mathbb{S}$$

provided

$$\mathbb{S} \in L^2(0, T; L^2(\Omega; R^{N^2})), \quad \mathbb{S} \text{ extended to be zero outside } \Omega,$$

where the operator $\nabla_x \Delta^{-1} \nabla_x$ is defined through (5.11).

Accordingly, if the viscous stress tensor $\mathbb{S}$ is given by (1.37), we get

$$(\nabla_x \Delta^{-1} \nabla_x) : \mathbb{S}$$

$$= \mathcal{F}_{\xi \to x}^{-1} \left[i \sum_{i,j=1}^{N} \frac{\xi_i \xi_j}{|\xi|^2} \left(\mu(\xi_j \mathcal{F}_{x \to \xi}[u^i] + \xi_i \mathcal{F}_{x \to \xi}[u^j]) + \lambda \delta_{i,j} \sum_{k=1}^{N} \xi_k \mathcal{F}_{x \to \xi}[u^k] \right) \right]$$

$$= (2\mu + \lambda) \text{div}_x \, \mathbf{u}$$

provided $\mathbf{u} \in L^2(I; W_0^{1,2}(\Omega; R^N))$, and $\mathbf{u}$ is extended to be zero outside Ω. Thus we can write

$$P_{\text{eff}} = p - (2\mu + \lambda) \, \text{div}_x \, \mathbf{u},$$

which is the standard definition used in the literature. Note that P_{eff} is nothing other than the amplitude of the normal viscous stress augmented by the hydrostatic pressure p, that is, the "real" pressure acting on a volume element of the fluid.

6.2 A result of P.-L. Lions on weak continuity

In the *compensated compactness* theory, *oscillations* in a weakly converging sequence of functions are studied through various relations between different weak limits of certain differential operators applied either to the sequence in question or to nonlinear compositions of functions in this sequence. Following this strategy, we consider sequences

$$\left\{ \begin{array}{c} \varrho_n \to \varrho \text{ weakly(-*) in } L^\infty(0, T; L^\gamma(\Omega)), \\ \mathbf{u}_n \to \mathbf{u} \text{ weakly in } L^2(0, T; W^{1,2}(\Omega; R^N)), \\ p_n \to p \text{ weakly in } L^r((0, T) \times \Omega) \text{ for a certain } r > 1, \end{array} \right\} \quad (6.3)$$

where ϱ_n, $\mathbf{u}_n$, and p_n are variational solutions of the system (1.34), (1.35) in the sense of Definitions 4.2, 4.3. At this stage, a concrete form of the constitutive equation for the pressure as well as for the temperature ϑ do not play any role, and we simply assume the pressure in (1.35) is represented by a function $p_n \in L^r((0, T) \times \Omega)$.

Moreover, we shall suppose that the kinetic energy

$$\varrho_n |\mathbf{u}_n|^2 \text{ is bounded in } L^\infty(0,T;L^1(\Omega)), \tag{6.4}$$

and

$$\mathbf{f}_n \to \mathbf{f} \text{ weakly(-*)} \quad \text{in } L^\infty((0,T)\times\Omega). \tag{6.5}$$

The following result states that the *effective viscous pressure* P_{eff} computed on the basis of ϱ_n and $\mathbf{u}_n$ behaves as though it was a strongly convergent quantity in the space $L^1((0,T)\times\Omega)$.

Proposition 6.1 *Let $\Omega \subset R^N$, $N \geq 2$, be an arbitrary domain. Assume that $\{\varrho_n\}_{n=1}^\infty$, $\{\mathbf{u}_n\}_{n=1}^\infty$, $\{p_n\}_{n=1}^\infty$, and $\{\mathbf{f}_n\}_{n=1}^\infty$ are sequences satisfying (6.3)–(6.5) with*

$$\gamma > \frac{N}{2}. \tag{6.6}$$

Let ϱ_n, $\mathbf{u}_n$ solve the continuity equation in the sense of renormalized solutions, that is, (4.13) holds in $\mathcal{D}'((0,T)\times\Omega)$ for any B,b satisfying (4.14), (4.15). Moreover, let ϱ_n, $\mathbf{u}_n$, p_n, and $\mathbf{f}_n$ solve the momentum equations (1.35) in $\mathcal{D}'((0,T)\times\Omega)$ with the stress tensor $\mathbb{S}_n$ given by the constitutive equation (1.37).

Then, passing to a subsequence if necessary, we have

$$\lim_{n\to\infty} \int_R \int_{R^N} \psi\eta \left(p_n - (2\mu+\lambda)\mathrm{div}_x\,\mathbf{u}_n\right) B(\varrho_n)\,\mathrm{d}\mathbf{x}\,\mathrm{d}t$$

$$= \int_R \int_{R^N} \psi\eta \left(p - (2\mu+\lambda)\mathrm{div}_x\,\mathbf{u}\right) \overline{B(\varrho)}\,\mathrm{d}\mathbf{x}\,\mathrm{d}t \tag{6.7}$$

for any $\psi \in \mathcal{D}(0,T)$, $\eta \in \mathcal{D}(\Omega)$, and any bounded continuous function B, where

$$B(\varrho_n) \to \overline{B(\varrho)} \text{ weakly(-*) in } L^\infty((0,T)\times\Omega).$$

Proposition 6.1 is a significant mathematical discovery of P.-L. Lions (see [85]). His original proof is based on the regularity properties of the *commutator*

$$u^i \mathcal{R}_i \mathcal{R}_j[\varrho u^i] - \mathcal{R}_i \mathcal{R}_j[\varrho u^i u^j],$$

where $\mathcal{R}$ is the *Riesz operator* introduced in (5.3). By virtue of the results of Coifman and Meyer [23], this quantity belongs to the Sobolev space $W^{1,q}$ provided $u^i \in W^{1,2}$ and $\varrho u^j \in L^r$ with $r > 2$ in which case $1/q = 1/r + 1/2$. Of course, this amounts to assuming $\varrho \in L^\gamma$ with $\gamma > 3$ if $N \geq 3$ which is much stronger than (6.6). However, a simple interpolation argument can be used to accommodate the general case $\gamma > N/2$. Here, we give an elementary proof of Proposition 6.1 based on the celebrated *Div–Curl Lemma*.

The relation (6.7) is a remarkable result because it brings to light an important property of the effective viscous pressure P_{eff} computed on the basis of ϱ_n, $\mathbf{u}_n$. It shows that this quantity enjoys some sort of weak continuity in the sense that its product with a weakly convergent sequence $\{B(\varrho_n)\}_{n=1}^{\infty}$ tends to the product of the corresponding weak limits. The information obtained becomes even more valuable once we realize that (6.7) contains the problematic term appearing on the right-hand side of (6.2).

Remark Even though we kept the original notation $(0, T)$ and Ω, one should realize that Proposition 6.1 is a *local* result. Accordingly we do not specify neither the boundary conditions satisfied by $\mathbf{u}$ nor the initial conditions for the density ϱ and the momentum $\varrho\mathbf{u}$. In the light of this observation, the hypothesis of boundedness of the pressure term p_n in L^r is fully justified in view of the local pressure estimates obtained in Chapter 5.

6.3 Weak continuity via compensated compactness

In this section, we give an elementary proof of Proposition 6.1. As already pointed out, this is a local result and, consequently, we are allowed to assume that $\Omega \subset R^N$ is a bounded domain. Our starting point is the following observation.

Lemma 6.1 *Let $\Omega \subset R^N$ be a domain. Let $\{\mathbf{v}_n\}_{n=1}^{\infty}$, $\{\mathbf{w}_n\}_{n=1}^{\infty}$ be two sequences of vector functions such that*

$$\mathbf{v}_n \rightarrow \mathbf{v} \text{ weakly in } L^p(\Omega; R^N), \quad \mathbf{w}_n \rightarrow \mathbf{w} \text{ weakly in } L^q(\Omega; R^N),$$

$$\frac{1}{p} + \frac{1}{q} \leq 1, \quad 1 < p, q < \infty.$$

Furthermore let

$$\operatorname{div}_x \mathbf{v}_n = 0 \text{ in } \mathcal{D}'(\Omega) \quad \text{for } n = 1, 2, \ldots$$

while

$$\mathbf{w}_n = \nabla_x G_n, \quad \text{with } G_n \text{ bounded in } W^{1,q}(\Omega).$$

Then

$$\mathbf{v}_n \cdot \mathbf{w}_n \rightarrow \mathbf{v} \cdot \mathbf{w} \quad \text{in } \mathcal{D}'(\Omega).$$

Remark The reader will have noticed that this is nothing other than a very simplified version of the celebrated *Div–Curl Lemma*—one of the major discoveries of the *compensated compactness* theory (cf. Theorem 11 and Example 3 of [111]).

Proof To begin with, we can secure a subsequence such that

$$G_n \to G \text{ weakly in } W^{1,q}(\Omega), \quad \text{where } \nabla_x G = \mathbf{w}.$$

On the other hand, taking an arbitrary test function $\eta \in \mathcal{D}(\Omega)$, we have

$$\int_\Omega \eta \, \mathbf{v}_n \cdot \mathbf{w}_n \, \mathrm{d}\mathbf{x} = \int_\Omega \eta \, \mathbf{v}_n \cdot \nabla_x G_n \, \mathrm{d}\mathbf{x} = - \int_\Omega G_n \mathbf{v}_n \cdot \nabla_x \eta \, \mathrm{d}\mathbf{x}$$

where the most right integral tends to the expression

$$- \int_\Omega G \mathbf{v} \cdot \nabla_x \eta \, \mathrm{d}\mathbf{x} = \int_\Omega \eta \, \mathbf{v} \cdot \mathbf{w} \, \mathrm{d}\mathbf{x}$$

as $n \to \infty$ because of compactness of the imbedding $W^{1,q}(K) \subset L^q(K)$ for any smooth bounded domain $K \subset \Omega$ (see Theorem 2.7). $\square$

Corollary 6.1 *Let $\Omega \subset R^N$ be an arbitrary domain.*
(i) *Let*

$$\mathbf{v}_n \to \mathbf{v} \text{ weakly in } L^p(\Omega; R^N), \quad \mathbf{w}_n \to \mathbf{w} \text{ weakly in } L^q(\Omega; R^N)$$

with

$$1 < p, \quad q < \infty, \quad \frac{1}{p} + \frac{1}{q} \leq 1.$$

Then

$$\mathbf{v}_n \cdot (\nabla_x \Delta^{-1} \mathrm{div}_x)[\mathbf{w}_n] - \mathbf{w}_n \cdot (\nabla_x \Delta^{-1} \mathrm{div}_x)[\mathbf{v}_n]$$
$$\to \mathbf{v} \cdot (\nabla_x \Delta^{-1} \mathrm{div}_x)[\mathbf{w}] - \mathbf{w} \cdot (\nabla_x \Delta^{-1} \mathrm{div}_x)[\mathbf{v}] \quad in \ \mathcal{D}'(\Omega).$$

(ii) *Under the same hypotheses, if*

$$B_n \to B \text{ weakly in } L^p(\Omega), \quad \mathbf{v}_n \to \mathbf{v} \text{ weakly in } L^q(\Omega; R^N),$$

then

$$(\nabla_x \Delta^{-1} \nabla_x)[B_n]\mathbf{v}_n - (\nabla_x \Delta^{-1} \mathrm{div}_x)[\mathbf{v}_n]B_n$$
$$\to (\nabla_x \Delta^{-1} \nabla_x)[B]\mathbf{v} - (\nabla_x \Delta^{-1} \mathrm{div}_x)[\mathbf{v}]B \text{ weakly in } \mathcal{D}'(\Omega; R^N).$$

Proof (i) By virtue of Lemma 5.1, the operators in question are bounded on the Lebesgue spaces L^r for any $1 < r < \infty$. One can write

$$\mathbf{v}_n \cdot (\nabla_x \Delta^{-1} \mathrm{div}_x)[\mathbf{w}_n] - \mathbf{w}_n \cdot (\nabla_x \Delta^{-1} \mathrm{div}_x)[\mathbf{v}_n]$$
$$= (\mathbf{v}_n - (\nabla_x \Delta^{-1} \mathrm{div}_x)[\mathbf{v}_n]) \cdot (\nabla_x \Delta^{-1} \mathrm{div}_x)[\mathbf{w}_n]$$
$$- (\mathbf{w}_n - (\nabla_x \Delta^{-1} \mathrm{div}_x)[\mathbf{w}_n]) \cdot (\nabla_x \Delta^{-1} \mathrm{div}_x)[\mathbf{v}_n].$$

Seeing that

$$\mathrm{div}_x(\mathbf{v}_n - (\nabla_x \Delta^{-1} \mathrm{div}_x)[\mathbf{v}_n]) = 0$$

we can use Lemma 6.1 to obtain

$$(\mathbf{v}_n - (\nabla_x \Delta^{-1} \mathrm{div}_x)[\mathbf{v}_n]) \cdot (\nabla_x \Delta^{-1} \mathrm{div}_x)[\mathbf{w}_n]$$
$$\to (\mathbf{v} - (\nabla_x \Delta^{-1} \mathrm{div}_x)[\mathbf{v}]) \cdot (\nabla_x \Delta^{-1} \mathrm{div}_x)[\mathbf{w}] \text{ in } \mathcal{D}'(\Omega).$$

Similarly,

$$(\mathbf{w}_n - (\nabla_x \Delta^{-1} \mathrm{div}_x)[\mathbf{w}_n]) \cdot (\nabla_x \Delta^{-1} \mathrm{div}_x)[\mathbf{v}_n]$$
$$\to (\mathbf{w} - (\nabla_x \Delta^{-1} \mathrm{div}_x)[\mathbf{w}]) \cdot (\nabla_x \Delta^{-1} \mathrm{div}_x)[\mathbf{v}]$$

which yields the desired conclusion.

(ii) Now, it is enough to observe that the i-th component of the vector

$$(\nabla_x \Delta^{-1} \nabla_x)[B_n] \mathbf{v}_n - (\nabla_x \Delta^{-1} \mathrm{div}_x)[\mathbf{v}_n] B_n$$

is equal to

$$(\nabla_x \Delta^{-1} \mathrm{div}_x)[B_n \mathbf{e}_i] \mathbf{v}_n - (\nabla_x \Delta^{-1} \mathrm{div}_x)[\mathbf{v}_n] B_n \mathbf{e}_i,$$

where $\{\mathbf{e}_1, \ldots, \mathbf{e}_N\}$ is the orthogonal basis of R^N. Consequently, the conclusion (ii) follows directly from (i).

$\square$

The second ingredient of the proof of Proposition 6.1 is the integral identity (5.10) obtained in Lemma 5.3. Taking $B = B(\varrho_n)$, $h = -b(\varrho_n) \mathrm{div}_x \mathbf{u}_n$, where B, b are bounded functions satisfying (4.14), (4.15), we get

$$\int_R \int_{R^N} \psi \eta (\xi \, p_n B(\varrho_n) - \mathbb{S}_n : (\nabla_x \Delta^{-1} \nabla_x)[\xi B(\varrho_n)]) \, \mathrm{dx} \, \mathrm{dt} = \sum_{j=1}^{7} I_j^n$$

$$+ \int_R \int_{R^N} \psi \mathbf{u}_n [\xi B(\varrho_n) (\nabla_x \Delta^{-1} \nabla_x) [\eta \varrho_n \mathbf{u}_n]$$
$$- (\nabla_x \Delta^{-1} \nabla_x) [\xi B(\varrho_n)] \eta \varrho_n \mathbf{u}_n]] \, \mathrm{dx} \, \mathrm{dt} \tag{6.8}$$

where

$$\mathbb{S}_n = \mu(\nabla_x \mathbf{u}_n + \nabla_x \mathbf{u}_n^t) + \lambda \, \mathbb{I} \, \mathrm{div}_x \mathbf{u}_n,$$

and

$$I_1^n = \int_R \int_{R^N} \psi \ (\mathbb{S}_n \nabla_x \eta) \cdot \mathcal{A}[\xi B(\varrho_n)] \, \mathrm{dx} \, \mathrm{dt},$$

$$I_2^n = -\int_R \int_{R^N} \psi \ p_n \ \nabla_x \eta \cdot \mathcal{A}[\xi B(\varrho_n)] \, \mathrm{dx} \, \mathrm{dt},$$

$$I_3^n = -\int_R \int_{R^N} \psi \eta \ \varrho_n \mathbf{f}_n \cdot \mathcal{A}[\xi B(\varrho_n)] \, \mathrm{dx} \, \mathrm{dt},$$

$$I_4^n = -\int_R \int_{R^N} \psi \ ([\varrho_n \mathbf{u}_n \otimes \mathbf{u}_n] \nabla_x \eta) \cdot \mathcal{A}[\xi B(\varrho_n)] \, \mathrm{dx} \, \mathrm{dt},$$

$$I_5^n = -\int_R \int_{R^N} \psi \eta \ \varrho_n \mathbf{u}_n \cdot \mathcal{A}[B(\varrho_n) \nabla_x \xi \cdot \mathbf{u}_n] \, \mathrm{dx} \, \mathrm{dt},$$

$$I_6^n = -\int_R \int_{R^N} \partial_t \psi \ \eta \ \varrho_n \mathbf{u}_n \cdot \mathcal{A}[\xi B(\varrho_n)] \, \mathrm{dx} \, \mathrm{dt},$$

$$I_7^n = \int_R \int_{R^N} \psi \eta \ \varrho_n \mathbf{u}_n \cdot \mathcal{A}[\xi b(\varrho_n) \mathrm{div}_x \mathbf{u}_n] \, \mathrm{dx} \, \mathrm{dt}$$

for any $\psi \in \mathcal{D}(0,T)$, η, $\xi \in \mathcal{D}(\Omega)$.

Now, by virtue of the *weak stability* property established in Lemma 4.6, the limit functions ϱ, $\mathbf{u}$, p, and f satisfy the momentum equation

$$\partial_t(\varrho \mathbf{u}) + \mathrm{div}_x(\varrho \mathbf{u} \otimes \mathbf{u}) + \nabla_x p = \mathrm{div}_x \mathbb{S} + \overline{\varrho \mathbf{f}} \qquad (6.9)$$

together with

$$\partial_t \overline{B(\varrho)} + \mathrm{div}_x \big(\overline{B(\varrho)} \mathbf{u}\big) + \overline{b(\varrho) \mathrm{div}_x \mathbf{u}} = 0 \qquad (6.10)$$

in $\mathcal{D}'((0,T) \times \Omega)$. Note that $\overline{B(\varrho) \mathbf{u}} = \overline{B(\varrho)} \mathbf{u}$ as

$$B(\varrho_n) \ \rightarrow \ \overline{B(\varrho)} \quad \text{in } C([0,T]; L_{\mathrm{weak}}^\gamma(\Omega)),$$

with $L^\gamma(\Omega)$ compactly imbedded in $W^{-1,2}(\Omega)$.

Thus another application of Lemma 5.3 yields

$$\int_R \int_{R^N} \psi \eta \Big(\xi p \overline{B(\varrho)} - \mathbb{S} : (\nabla_x \Delta^{-1} \nabla_x) \big[\xi \overline{B(\varrho)}\big] \Big) \, \mathrm{dx} \, \mathrm{dt} = \sum_{j=1}^{7} I_j$$

$$+ \int_R \int_{R^N} \psi \mathbf{u} \Big[\xi \overline{B(\varrho)} (\nabla_x \Delta^{-1} \nabla_x)[\eta \varrho \mathbf{u}] - (\nabla_x \Delta^{-1} \nabla_x)[\xi \overline{B(\varrho)}] \eta \varrho \mathbf{u} \Big] \, \mathrm{dx} \, \mathrm{dt},$$

$$\qquad (6.11)$$

where

$$\mathbb{S} = \mu(\nabla_x \mathbf{u} + \nabla_x \mathbf{u}^t) + \lambda \ \mathbb{I} \ \mathrm{div}_x \mathbf{u},$$

and

$$I_1 = \int_R \int_{R^N} \psi \, (\mathbb{S}\nabla_x\eta) \cdot \mathcal{A}\big[\xi\overline{B(\varrho)}\big] \, \mathrm{d}\mathbf{x}\,\mathrm{d}t,$$

$$I_2 = -\int_R \int_{R^N} \psi \, p \, \nabla_x\eta \cdot \mathcal{A}\big[\xi\overline{B(\varrho)}\big] \, \mathrm{d}\mathbf{x}\,\mathrm{d}t,$$

$$I_3 = -\int_R \int_{R^N} \psi\eta \, \overline{\varrho\mathbf{f}} \cdot \mathcal{A}\big[\xi\overline{B(\varrho)}\big] \, \mathrm{d}\mathbf{x}\,\mathrm{d}t,$$

$$I_4 = -\int_R \int_{R^N} \psi \, ([\varrho\mathbf{u}\otimes\mathbf{u}]\nabla_x\eta) \cdot \mathcal{A}\big[\xi\overline{B(\varrho)}\big] \, \mathrm{d}\mathbf{x}\,\mathrm{d}t,$$

$$I_5 = -\int_R \int_{R^N} \psi\eta \, \varrho\mathbf{u} \cdot \mathcal{A}\big[\overline{B(\varrho)}\nabla_x\xi \cdot \mathbf{u}\big] \, \mathrm{d}\mathbf{x}\,\mathrm{d}t,$$

$$I_6 = -\int_R \int_{R^N} \partial_t\psi \, \eta \, \varrho\mathbf{u} \cdot \mathcal{A}\big[\xi\overline{B(\varrho)}\big] \, \mathrm{d}\mathbf{x}\,\mathrm{d}t,$$

$$I_7 = \int_R \int_{R^N} \psi\eta \, \varrho\mathbf{u} \cdot \mathcal{A}\big[\xi\overline{b(\varrho)\mathrm{div}_x\,\mathbf{u}}\big] \, \mathrm{d}\mathbf{x}\,\mathrm{d}t.$$

The singular integral operator $\mathcal{A}$ introduced in (5.4) enjoys the smoothing properties specified in Lemma 5.2. In particular, since $B(\varrho_n)$ satisfies the renormalized continuity equation (4.13), one has

$$B(\varrho_n) \to \overline{B(\varrho)} \quad \text{in } C([0,T]; L^p_{\mathrm{weak}}(\Omega)) \quad \text{for any finite } p \geq 1, \qquad (6.12)$$

and, consequently,

$$\mathcal{A}[\xi B(\varrho_n)] \to \mathcal{A}\big[\xi\overline{B(\varrho)}\big] \quad \text{in } C(K)$$

for any compact $K \subset (0,T) \times \Omega$. Thus

$$I_j^n \to I_j \quad \text{as } n \to \infty \quad \text{for } j = 1,2,3; \qquad (6.13)$$

and, by virtue of Lemma 4.6,

$$I_4^n \to I_4 \quad \text{and} \quad I_6^n \to I_6 \quad \text{as } n \to \infty. \qquad (6.14)$$

Note that the convective terms $\varrho_n\mathbf{u}_n \otimes \mathbf{u}_n$ are bounded in the Lebesgue space $L^2(0,T; L^{c_2}(\Omega; R^{N^2}))$ as in Theorem 3.1.

Next, we have

$$\left\{ \begin{array}{l} B(\varrho_n)\nabla_x\xi \cdot \mathbf{u}_n \to \overline{B(\varrho)}\nabla_x\xi \cdot \mathbf{u}, \\[2mm] b(\varrho_n)\,\mathrm{div}_x\,\mathbf{u}_n \to \overline{b(\varrho)\,\mathrm{div}_x\,\mathbf{u}} \end{array} \right\} \quad \text{weakly in } L^2((0,T)\times\Omega);$$

whence by Lemma 5.2,

$$\mathcal{A}[B(\varrho_n)\nabla_x\xi \cdot \mathbf{u}_n] \to \mathcal{A}\big[\overline{B(\varrho)}\nabla_x\xi \cdot \mathbf{u}\big],$$

and

$$A[b(\varrho_n) \operatorname{div}_x \mathbf{u}_n] \to A\overline{[b(\varrho) \operatorname{div}_x \mathbf{u}]}$$

weakly in $L^2(0, T; W^{1,2}(\Omega; R^N))$.

Similarly as in (4.30),

$$\varrho_n \mathbf{u}_n \to \varrho \mathbf{u} \text{ in } C([0, T]; L^{m_\infty}_{\text{weak}}(\Omega; R^N)), \quad m_\infty = \frac{2\gamma}{\gamma + 1}, \tag{6.15}$$

and we infer

$$I_5^n \to I_5, \quad I_7^n \to I_7 \text{ as } n \to \infty. \tag{6.16}$$

Finally, the relations (6.12) and (6.15) together with Corollary 6.1 imply

$$(\nabla_x \Delta^{-1} \nabla_x)[\xi B(\varrho_n(t))] \, \eta \varrho_n \mathbf{u}_n(t) - \xi B(\varrho_n(t)) \, (\nabla_x \Delta^{-1} \operatorname{div}_x)[\eta \varrho_n \mathbf{u}_n(t)]$$
$$\to (\nabla_x \Delta^{-1} \nabla_x)[\xi \overline{B(\varrho)}(t)] \, \eta \varrho \mathbf{u}(t) - \xi \overline{B(\varrho)}(t) \, (\nabla_x \Delta^{-1} \operatorname{div}_x)[\eta \varrho \mathbf{u}(t)]$$

weakly in $L^r(\Omega; R^N)$ for $1 \le r < \frac{2\gamma}{\gamma + 1}$

for each fixed $t \in [0, T]$. As $\gamma > N/2$, one can apply Theorem 2.8 to obtain

$$(\nabla_x \Delta^{-1} \nabla_x)[\xi B(\varrho_n)] \, \eta \varrho_n \mathbf{u}_n - \xi B(\varrho_n) \, (\nabla_x \Delta^{-1} \operatorname{div}_x)[\eta \varrho_n \mathbf{u}_n]$$
$$\to (\nabla_x \Delta^{-1} \nabla_x)[\xi \overline{B(\varrho)}] \, \eta \varrho \mathbf{u} - \xi \overline{B(\varrho)} \, (\nabla_x \Delta^{-1} \operatorname{div}_x)[\eta \varrho \mathbf{u}]$$
(strongly) in $L^2(0, T; W^{-1,2}(\Omega; R^N))$

and, consequently,

$$\int_R \int_{R^N} \psi \mathbf{u}_n \cdot \left[(\nabla_x \Delta^{-1} \nabla_x)[\xi B(\varrho_n)] \eta \varrho_n \mathbf{u}_n - \xi B(\varrho_n)(\nabla_x \Delta^{-1} \operatorname{div}_x)[\eta \varrho_n \mathbf{u}_n] \right] \mathrm{dx}\,\mathrm{dt}$$
$$\to \int_R \int_{R^N} \psi \mathbf{u} \cdot \left[(\nabla_x \Delta^{-1} \nabla_x)[\xi \overline{B(\varrho)}] \, \eta \varrho \mathbf{u} - \xi \overline{B(\varrho)} \, (\nabla_x \Delta^{-1} \operatorname{div}_x)[\eta \varrho \mathbf{u}] \right] \mathrm{dx}\,\mathrm{dt}$$

which, together with (6.13), (6.14), and (6.16), yields

$$\lim_{n \to \infty} \int_R \int_{R^N} \psi \eta \Big(\xi \, p_n B(\varrho_n) - \mathbb{S}_n : (\nabla_x \Delta^{-1} \nabla_x)[\xi B(\varrho_n)] \Big) \mathrm{dx}\,\mathrm{dt}$$
$$= \int_R \int_{R^N} \psi \eta \Big(\xi \, p \overline{B(\varrho)} - \mathbb{S} : (\nabla_x \Delta^{-1} \nabla_x)[\xi \overline{B(\varrho)}] \Big) \mathrm{dx}\,\mathrm{dt} \tag{6.17}$$

for any $\psi \in \mathcal{D}(0, T)$, $\eta, \xi \in \mathcal{D}(\Omega)$.

To conclude, we compute

$$\int_R \int_{R^N} \psi \eta \, \mathbb{S}_n : (\nabla_x \Delta^{-1} \nabla_x)[\xi B(\varrho_n)] \, \mathrm{d}\mathbf{x} \, \mathrm{d}t$$

$$= \int_R \int_{R^N} \psi \xi \, (\nabla_x \Delta^{-1} \nabla_x) : (\eta \mathbb{S}_n) B(\varrho_n) \, \mathrm{d}\mathbf{x} \, \mathrm{d}t$$

$$= \int_R \int_{R^N} \psi \xi \, (2\mu + \lambda) \, \mathrm{div}_x(\eta \mathbf{u}_n) B(\varrho_n) \, \mathrm{d}\mathbf{x} \, \mathrm{d}t$$

$$- \int_R \int_{R^N} \psi \xi B(\varrho_n) \big[2\mu (\nabla_x \Delta^{-1} \nabla_x) : (\mathbf{u}_n \otimes \nabla_x \eta) + \lambda \mathbf{u}_n \cdot \nabla_x \eta \big] \, \mathrm{d}\mathbf{x} \, \mathrm{d}t$$

$$= \int_R \int_{R^N} \psi \xi \eta \, (2\mu + \lambda) \, \mathrm{div}_x \, \mathbf{u}_n \, B(\varrho_n) \, \mathrm{d}\mathbf{x} \, \mathrm{d}t$$

$$+ \int_R \int_{R^N} 2\mu \, \psi \eta \, B(\varrho_n) \big[(\nabla_x \Delta^{-1} \nabla_x) : (\mathbf{u}_n \otimes \nabla_x \eta) + \mathbf{u}_n \cdot \nabla_x \eta \big] \, \mathrm{d}\mathbf{x} \, \mathrm{d}t; \tag{6.18}$$

and, similarly,

$$\int_R \int_{R^N} \psi \eta \, \mathbb{S} : (\nabla_x \Delta^{-1} \nabla_x) \big[\xi \overline{B(\varrho)} \big] \, \mathrm{d}\mathbf{x} \, \mathrm{d}t$$

$$= \int_R \int_{R^N} \psi \xi \eta \, (2\mu + \lambda) \, \mathrm{div}_x \, \mathbf{u} \, \overline{B(\varrho)} \, \mathrm{d}\mathbf{x} \, \mathrm{d}t$$

$$+ \int_R \int_{R^N} 2\mu \, \psi \eta \, \overline{B(\varrho)} \big[(\nabla_x \Delta^{-1} \nabla_x) : (\mathbf{u} \otimes \nabla_x \eta) + \mathbf{u} \cdot \nabla_x \eta \big] \, \mathrm{d}\mathbf{x} \, \mathrm{d}t. \tag{6.19}$$

Taking (6.12) into account one can see that the relation (6.17) together with (6.18), (6.19) imply (6.7) for any bounded function B satisfying (4.15).

If B is a general bounded function, there exists a sequence $\{B_m\}_{m=1}^{\infty}$ of functions satisfying (4.15) which are bounded uniformly with respect to m and such that $B_m \to B - B(0)$ on compacts in R. Since both the pressure p_n and $\mathrm{div}_x \, \mathbf{u}_n$ belong to a bounded set in $L^r((0, T) \times \Omega)$, $r > 1$, we can pass to the limit in (6.7) for $m \to \infty$ to obtain the same relation for the limit function B. This completes the proof of Proposition 6.1.

6.4 The oscillations defect measure

In this section we broaden our previous analysis to describe the possible *density oscillations* discussed above. As we have seen, a suitable tool to do this is a "*defect measure*" $\overline{B(\varrho)} - B(\varrho)$, the time evolution of which is governed by equation (6.2) provided we can show that the limit density ϱ satisfies the renormalized continuity equation (4.13). As a typical velocity field $\mathbf{u}$ we consider here belongs to the space $L^2(0, T; W_0^{1,2}(\Omega; R^N))$, we can use the regularizing operators as in Section 4.1 to conclude that a distributional solution $\varrho, \mathbf{u}$ of (1.34) will satisfy also (4.13) if

$$\varrho \in L^2(0, T; L^2(\Omega)) \tag{6.20}$$

(see Corollary 4.1).

In view of the pressure estimates obtained in Section 4.3, relation (6.20) holds at least locally on $(0, T) \times \Omega$ provided

$$\gamma \geq \frac{3N}{N+2}$$

(see Proposition 5.1). However, such a condition would mean an unnecessary technical restriction when compared with the "optimal value" of the "adiabatic" exponent $\gamma > N/2$.

As a matter of fact, condition (6.20) is not optimal. Clearly, one obtains the same conclusion when, for instance, ϱ_n are renormalized solutions, and

$$\varrho_n \to \varrho \text{ (strongly)} \quad \text{in } L^1((0, T) \times \Omega),$$

as a direct consequence of the Lebesgue dominated convergence theorem. Of course, both phenomena may occur simultaneously in the sense that ϱ_n are renormalized solutions of (1.34), and

- ϱ_n converges to ϱ strongly in $L^1(O)$;
- ϱ_n are bounded in $L^2(((0, T) \times \Omega) \setminus O)$ for a certain $O \subset ((0, T) \times \Omega)$.

Both possibilities are captured by a more general concept of describing the density oscillations in terms of a new quantity called *oscillations defect measure* $\mathbf{osc}_p[\varrho_n \to \varrho]$. For a sequence

$$\varrho_n \to \varrho \text{ weakly in } L^1(O),$$

we set

$$\mathbf{osc}_p[\varrho_n \to \varrho](O) \equiv \sup_{k \geq 1} \left(\limsup_{n \to \infty} \int_O |T_k(\varrho_n) - T_k(\varrho)|^p \, \mathrm{d}\mathbf{x} \, \mathrm{d}t \right), \tag{6.21}$$

where T_k are the cut-off functions introduced in (2.12).

We claim the following result.

Proposition 6.2 *Let $\Omega \subset R^N$, $N \geq 2$ be an arbitrary domain. Moreover, let $\{\varrho_n\}_{n=1}^\infty$, $\{\mathbf{u}_n\}_{n=1}^\infty$, $\{p_n\}_{n=1}^\infty$, and $\{\mathbf{f}_n\}_{n=1}^\infty$ be sequences satisfying (6.3–6.5), with*

$$\gamma > \frac{N}{2},$$

where, in addition, ϱ_n, $\mathbf{u}_n$ solve the renormalized continuity equation (4.13) in $\mathcal{D}'((0, T) \times \Omega)$, and ϱ_n, $\mathbf{u}_n$, p_n, and $\mathbf{f}_n$ solve the momentum equations (1.35) in

$\mathcal{D}'((0,T) \times \Omega)$ *with the stress tensor* $\mathbb{S}_n$ *given by the constitutive equation (1.37).* *Finally, suppose that* $\varrho_n \geq 0$, *and that the pressure* p_n *can be written as*

$$p_n = p_c(\varrho_n) + p_m(\varrho_n) + p_b^n, \quad n = 1, 2, \ldots, \tag{6.22}$$

where

$$p_c : [0, \infty) \to R \text{ is a convex function}, \quad p_c(0) = 0, \quad p_c(\varrho) \geq a\varrho^\gamma$$

for all $\varrho \geq 0$ *and a certain* $a > 0$;

$$p_m : [0, \infty) \to R \text{ is a non-decreasing function};$$

and

$$p_b^n \text{ are bounded in } L^r((0,T) \times \Omega) \quad \text{with } r = \frac{\gamma+1}{\gamma} = (\gamma+1)'.$$

Then

$$\operatorname{osc}_{\gamma+1}[\varrho_n \to \varrho](O) \leq c(|O|) < \infty$$

for any bounded $O \subset ((0,T) \times \Omega)$.

Remark Hypothesis (6.22) allows for perturbations of the *elastic pressure* component of the form

$$|p_e(\varrho) - a\varrho^\gamma| \leq c(1 + \varrho^{\gamma-1}),$$

in particular, the pressure need not be monotone. Indeed if $(0,T) \times \Omega$ is bounded and $\{\varrho_n\}_{n=1}^\infty$ is a bounded sequence in $L^\infty(0,T; L^\gamma(\Omega))$, then

$$1 + \varrho_n^{\gamma-1} \text{ is bounded in } L^{(\gamma+1)/\gamma}((0,T) \times \Omega)$$

as required.

Remark In accordance with the *a priori* estimates (3.36), the *thermal pressure* component $\vartheta_n p_\vartheta(\varrho_n)$ is always bounded in $L^2((0,T) \times \Omega)$ independently of n, and, consequently, it can be included in the bounded term p_b^n.

Proof As the result is local, we can assume $O = ((0,T) \times \Omega)$ where $\Omega \subset R^N$ is a bounded domain. Moreover, in accordance with our hypotheses, one can assume $p_m(0) = 0$,

$$p_c(\varrho_n) \to \overline{p_c(\varrho)} \text{ weakly in } L^r((0,T) \times \Omega),$$

and

$$p_m(\varrho_n) \to \overline{p_m(\varrho)} \text{ weakly in } L^r((0,T) \times \Omega),$$

where $r > 1$ is the same as in (6.3).

Furthermore, we have

$$p_b^n \to p_b \text{ weakly in } L^{(\gamma+1)/\gamma}((0,T) \times \Omega)$$

passing to a subsequence as the case may be.

(i) Now, we can apply Proposition 6.1 to obtain

$$\lim_{n\to\infty} \int_0^T \int_\Omega p_{\mathrm c}(\varrho_n)\, T_k(\varrho_n) - \overline{p_{\mathrm c}(\varrho)}\; \overline{T_k(\varrho)}\,\mathrm{d}\mathbf{x}\,\mathrm{d}t$$

$$+ \lim_{n\to\infty} \int_0^T \int_\Omega p_{\mathrm m}(\varrho_n)\, T_k(\varrho_n) - \overline{p_{\mathrm m}(\varrho)}\; \overline{T_k(\varrho)}\,\mathrm{d}\mathbf{x}\,\mathrm{d}t$$

$$= (2\mu+\lambda)\lim_{n\to\infty} \int_0^T \int_\Omega \operatorname{div}_x \mathbf{u}_n T_k(\varrho_n) - \operatorname{div}_x \mathbf{u}\, \overline{T_k(\varrho)}\,\mathrm{d}\mathbf{x}\,\mathrm{d}t$$

$$+ \lim_{n\to\infty} \int_0^T \int_\Omega p_{\mathrm b}\, \overline{T_k(\varrho)} - p_{\mathrm b}^n\, T_k(\varrho_n)\,\mathrm{d}\mathbf{x}\,\mathrm{d}t \tag{6.23}$$

where

$$T_k(\varrho_n) \to \overline{T_k(\varrho)} \text{ weakly(-*) in } L^\infty((0,T)\times\Omega).$$

We have

$$\lim_{n\to\infty} \int_0^T \int_\Omega p_{\mathrm c}(\varrho_n)\, T_k(\varrho_n) - \overline{p_{\mathrm c}(\varrho)}\; \overline{T_k(\varrho)}\,\mathrm{d}\mathbf{x}\,\mathrm{d}t$$

$$= \lim_{n\to\infty} \int_0^T \int_\Omega \big(p_{\mathrm c}(\varrho_n) - p_{\mathrm c}(\varrho)\big)\big(T_k(\varrho_n) - T_k(\varrho)\big)\,\mathrm{d}\mathbf{x}\,\mathrm{d}t$$

$$+ \int_0^T \int_\Omega \big(\overline{p_{\mathrm c}(\varrho)} - p_{\mathrm c}(\varrho)\big)\big(T_k(\varrho) - \overline{T_k(\varrho)}\big)\,\mathrm{d}\mathbf{x}\,\mathrm{d}t$$

$$\geq \lim_{n\to\infty} \int_0^T \int_\Omega \big(p_{\mathrm c}(\varrho_n) - p_{\mathrm c}(\varrho)\big)\big(T_k(\varrho_n) - T_k(\varrho)\big)\,\mathrm{d}\mathbf{x}\,\mathrm{d}t \tag{6.24}$$

as $p_{\mathrm c}$ is convex and T_k concave (see Theorem 2.11).

By virtue of hypothesis (6.22), the function $p_{\mathrm c}$ is strictly increasing convex on $[0,\infty)$. Consequently,

$$p_{\mathrm c}(y) - p_{\mathrm c}(z) = \int_z^y p_{\mathrm c}'(s)\,\mathrm{d}s \geq \int_z^y p_{\mathrm c}'(s-z)\,\mathrm{d}s = p_{\mathrm c}(y-z) \quad \text{for all } y \geq z \geq 0,$$

and

$$p_{\mathrm c}(|T_k(y) - T_k(z)|) \leq p_{\mathrm c}(|y - z|).$$

Thus, by virtue of (6.22),

$$a|T_k(\varrho_n) - T_k(\varrho)|^{\gamma+1} \leq p_{\mathrm c}(|T_k(\varrho_n) - T_k(\varrho)|)|T_k(\varrho_n) - T_k(\varrho)|$$

$$\leq p_{\mathrm c}(|\varrho_n - \varrho|)|T_k(\varrho_n) - T_k(\varrho)|$$

$$\leq \big(p_{\mathrm c}(\varrho_n) - p_{\mathrm c}(\varrho)\big)\big(T_k(\varrho_n) - T_k(\varrho)\big). \tag{6.25}$$

Relation (6.24) together with (6.25) yield

$$\lim_{n\to\infty} \int_0^T \int_\Omega p_c(\varrho_n)\, T_k(\varrho_n) - \overline{p_c(\varrho)}\; \overline{T_k(\varrho)}\, d\mathbf{x}\, dt$$

$$\geq a \limsup_{n\to\infty} \int_0^T \int_\Omega |T_k(\varrho_n) - T_k(\varrho)|^{\gamma+1}\, d\mathbf{x}\, dt. \qquad (6.26)$$

(ii) As the next step, we have

$$\left(T_k(\varrho_n) - T_k((p_{\mathrm{m}} + \varepsilon\mathrm{Id})^{-1}\big(\overline{p_{\mathrm{m}}(\varrho)} + \varepsilon\varrho)\big)\right)\left(p_{\mathrm{m}}(\varrho_n) - \overline{p_{\mathrm{m}}(\varrho)} + \varepsilon(\varrho_n - \varrho)\right) \geq 0$$

for any $\varepsilon > 0$ as the function $\varrho \mapsto p_{\mathrm{m}}(\varrho) + \varepsilon\varrho$ is invertible strictly increasing on $[0, \infty)$.

Thus integration over $(0, T) \times \Omega$ and passing to limit for $n \to \infty$ yields

$$\lim_{n\to\infty} \int_0^T \int_\Omega p_{\mathrm{m}}(\varrho_n) T_k(\varrho_n) - \overline{p_{\mathrm{m}}(\varrho)}\; \overline{T_k(\varrho)}\, d\mathbf{x}\, dt$$

$$\geq -4\varepsilon k \sup_{n\geq 1} \|\varrho_n\|_{L^1((0,T)\times\Omega)}. \qquad (6.27)$$

Since $\varepsilon > 0$ can be taken arbitrarily small, we infer

$$\lim_{n\to\infty} \int_0^T \int_\Omega p_{\mathrm{m}}(\varrho_n) T_k(\varrho_n) - \overline{p_{\mathrm{m}}(\varrho)}\; \overline{T_k(\varrho)}\, d\mathbf{x}\, dt \geq 0. \qquad (6.28)$$

(iii) In order to conclude, we deduce

$$\lim_{n\to\infty} \int_0^T \int_\Omega p_{\mathrm{b}}\, \overline{T_k(\varrho)} - p_{\mathrm{b}}^n\, T_k(\varrho_n)\, d\mathbf{x}\, dt$$

$$\leq \int_0^T \int_\Omega p_{\mathrm{b}}|\overline{T_k(\varrho)} - T_k(\varrho)|\, d\mathbf{x}\, dt + \limsup_{n\to\infty} \int_0^T \int_\Omega p_{\mathrm{b}}^n |T_k(\varrho_n) - T_k(\varrho)|\, d\mathbf{x}\, dt$$

$$\leq 2 \sup_{n\geq 1} \|p_{\mathrm{b}}^n\|_{L^{(\gamma+1)/\gamma}((0,T)\times\Omega)} \limsup_{n\to\infty} \|T_k(\varrho_n) - T_k(\varrho)\|_{L^{\gamma+1}((0,T)\times\Omega)}. \qquad (6.29)$$

Similarly, we estimate

$$\lim_{n\to\infty} \int_0^T \int_\Omega \mathrm{div}_x\, \mathbf{u}_n T_k(\varrho_n) - \mathrm{div}_x\, \mathbf{u}\, \overline{T_k(\varrho)}\, d\mathbf{x}\, dt$$

$$\leq 2 \sup_{n\geq 1} \|\mathrm{div}_x\, \mathbf{u}\|_{L^2((0,T)\times\Omega)} \lim_{n\to\infty} \sup \|T_k(\varrho_n) - T_k(\varrho)\|_{L^2((0,T)\times\Omega)}. \qquad (6.30)$$

Since we assumed that $O = (0, T) \times \Omega$ is bounded, relations (6.26)–(6.30) together with Hölder's inequality (2.1) yield the conclusion of Proposition 6.2.

□

6.5 Renormalized solutions revisited

In Section 4.1, we introduced and examined in detail the *renormalized solutions* of the continuity equation (1.34). In particular, we proved the following result (see Corollary 4.1): if the density $\varrho \in L^2(0,T; L^2(\Omega))$ and $\mathbf{u} \in L^2(0,T; W^{1,2}(\Omega; R^N))$ solve (1.34) in the sense of distributions, then ϱ is also a renormalized solution. Unfortunately, however, the best possible *a priori* estimate on ϱ established in Proposition 5.1 does not always guarantee the square integrability while the L^2-estimates of the gradient of the velocity seem to be optimal in view of the available mathematical techniques.

It is desirable to attempt an answer whether or not given functions ϱ, $\mathbf{u}$ represent a renormalized solution of (1.34) without using the integrability of ϱ. The next assertion formulates a sufficient condition in terms of the *oscillations defect measure*.

Proposition 6.3   *Let $\Omega \subset R^N$, $N \geq 2$ be an arbitrary domain. Let $\{\varrho_n\}_{n=1}^{\infty}$ be a sequence of non-negative functions such that*

$$\varrho_n \to \varrho \ weakly(\text{-}*) \ in \ L^{\infty}(0,T; L^{\gamma}(\Omega)), \quad \gamma > \frac{2N}{N+2},$$

and

$$\mathbf{osc}_p[\varrho_n \to \varrho](O) < c(O) \quad for \ some \ p > 2$$

for any bounded $O \subset (0,T) \times \Omega$.
Let, moreover,

$$\mathbf{u}_n \to \mathbf{u} \ weakly \ in \ L^2(0,T; W^{1,2}(\Omega; R^N))$$

where ϱ_n, $\mathbf{u}_n$ solve (1.34) in the sense of renormalized solutions on $(0,T) \times \Omega$ (see Definition 4.1).
Then ϱ, $\mathbf{u}$ is a renormalized solution of (1.34) on $(0,T) \times \Omega$ in the sense of Definition 4.1.

Proof   (i) This is a local result and, consequently, we are allowed to assume that Ω is bounded. Since ϱ_n, $\mathbf{u}_n$ solve (1.34) in $\mathcal{D}'((0,T) \times \Omega)$, one can use Corollary 2.1 to deduce

$$\varrho_n \to \varrho \quad in \ C([0,T]; L^{\gamma}_{\text{weak}}(\Omega)).$$

By virtue of Theorem 2.8, the space L^{γ} is compactly imbedded in $W^{-1,2}$ and, consequently,

$$\varrho_n \mathbf{u}_n \to \varrho \mathbf{u} \ weakly \ in \ L^2(0,T; L^{m_2}(\Omega; R^N)), \quad with \ m_2 = \frac{2N\gamma}{2N + \gamma(N-2)}.$$

Accordingly, the limit functions ϱ, $\mathbf{u}$ solve the equation of continuity (1.34) in the sense of distributions on $(0, T) \times \Omega$.

(ii) Since ϱ_n is a renormalized solution of (1.34), we get

$$T_k(\varrho_n) \to \overline{T_k(\varrho)} \quad \text{weakly(-*) in } L^\infty((0, T) \times \Omega),$$

and

$$T_k(\varrho_n)\mathbf{u}_n \to \overline{T_k(\varrho)}\mathbf{u} \quad \text{weakly in } L^2((0, T); L^{2^*}(\Omega; R^N))$$

passing to subsequences as the case may be. Here, T_k are the cut-off functions introduced in (2.11), and the critical Sobolev exponent 2^* is determined by (5.9).

Furthermore, we can assume

$$(T_k'(\varrho_n)\varrho_n - T_k(\varrho_n))\text{div}_x\, \mathbf{u}_n \to \overline{(T_k'(\varrho)\varrho - T_k(\varrho))\text{div}_x\, \mathbf{u}}$$

$$\text{weakly in } L^2((0, T) \times \Omega),$$

and, since ϱ_n, $\mathbf{u}_n$ are renormalized solutions of (1.34),

$$\partial_t \overline{T_k(\varrho)} + \text{div}_x\big(\overline{T_k(\varrho)}\mathbf{u}\big) + \overline{(T_k'(\varrho)\varrho - T_k(\varrho))\text{div}_x\, \mathbf{u}} = 0 \quad \text{in } \mathcal{D}'((0, T) \times \Omega). \quad (6.31)$$

(iii) At this stage, it is possible to apply the regularizing operators $v \mapsto [v]_x^\varepsilon$ to both sides of (6.31) or, more precisely, to use Proposition 4.2 to obtain

$$\partial_t B(\overline{T_k(\varrho)}) + \text{div}_x(B(\overline{T_k(\varrho)})\mathbf{u}) + b(\overline{T_k(\varrho)})\,\text{div}_x\mathbf{u}$$

$$= B'(\overline{T_k(\varrho)})\,\overline{(T_k(\varrho) - T_k'(\varrho)\varrho)\text{div}_x\, \mathbf{u}} \quad \text{in } \mathcal{D}'((0, T) \times \Omega). \quad (6.32)$$

for any bounded $b \in C^1[0, \infty)$ and B as in (4.14), (4.15).

Since we already know that ϱ, $\mathbf{u}$ is a distributional solution of (1.34), it is enough to show that (4.13) holds for any

$$B \in C^1[0, \infty), \quad B'(z) = 0 \text{ for all } z \geq z_B, \quad b(z) = B'(z)z - B(z). \quad (6.33)$$

Indeed for general B, b satisfying (4.15), (4.16) the same result follows from the Lebesgue convergence theorem and a simple approximation argument.

A natural way to achieve this is, of course, to let $k \to \infty$ in (6.32). Utilizing the weak lower semi-continuity of the norm we deduce

$$\|\overline{T_k(\varrho)} - \varrho\|_{L^1((0,T)\times\Omega)} \leq \liminf_{n\to\infty} \|T_k(\varrho_n) - \varrho_n\|_{L^1((0,T)\times\Omega)}$$

$$\leq \sup_{n\geq 1} \int_{\{\varrho_n \geq k\}} \varrho_n \, d\mathbf{x}\, dt \leq k^{1-\gamma} \sup_{n\geq 1} \|\varrho_n\|_{L^\gamma((0,T)\times\Omega)},$$

where the right-hand side tends to zero for $k \to \infty$. Consequently, we have

$$B(\overline{T_k(\varrho)}) \to B(\varrho), \quad b(\overline{T_k(\varrho)}) \to b(\varrho) \quad \text{in } L^r((0, T) \times \Omega) \quad \text{for any } r \geq 1. \quad (6.34)$$

(iv) In order to complete the proof, we have to show that the right-hand side of (6.32) tends to zero for large k. To this end, we estimate

$$\left\| B'(\overline{T_k(\varrho)}) \; \overline{(T_k(\varrho) - T_k'(\varrho)\varrho)\mathrm{div}_x\, \mathbf{u}} \right\|_{L^1((0,T)\times\Omega)}$$

$$\leq \max_{z\geq 0}|B'(z)| \int_{\{\overline{T_k(\varrho)}\leq z_B\}} \left| \overline{(T_k(\varrho) - T_k'(\varrho)\varrho)\mathrm{div}_x\, \mathbf{u}} \right| \mathrm{d}\mathbf{x}\,\mathrm{d}t$$

$$\leq \max_{z\geq 0}|B'(z)| \sup_{n\geq 0}\|\mathrm{div}_x\, \mathbf{u}_n\|_{L^2((0,T)\times\Omega)}$$

$$\times \liminf_{n\to\infty}\|T_k(\varrho_n) - T_k'(\varrho_n)\varrho_n\|_{L^2(\{\overline{T_k(\varrho)}\leq z_B\})}. \tag{6.35}$$

By virtue of the interpolation inequality (2.2), the last term can be treated as

$$\|T_k(\varrho_n) - T_k'(\varrho_n)\varrho_n\|_{L^2(\{\overline{T_k(\varrho)}\leq z_B\})}$$

$$\leq \|T_k(\varrho_n) - T_k'(\varrho_n)\varrho_n\|_{L^1((0,T)\times\Omega)}^{\omega}$$

$$\times \|T_k(\varrho_n) - T_k'(\varrho_n)\varrho_n\|_{L^{\gamma+1}(\{\overline{T_k(\varrho)}\leq z_B\})}^{1-\omega} \quad \text{for a certain } \omega \in (0,1),$$

where, similarly as above,

$$\|T_k(\varrho_n) - T_k'(\varrho_n)\varrho_n\|_{L^1((0,T)\times\Omega)}$$

$$\leq 2\int_{\{\varrho_n\geq k\}} \varrho_n\,\mathrm{d}\mathbf{x}\,\mathrm{d}t \leq 2k^{1-\gamma}\sup_{n\geq 1}\|\varrho_n\|_{L^\infty(0,T;L^\gamma(\Omega))}. \tag{6.36}$$

In order to conclude, we compute

$$\limsup_{n\to\infty}\|T_k(\varrho_n) - T_k'(\varrho_n)\varrho_n\|_{L^{\gamma+1}(\{\overline{T_k(\varrho)}\leq z_B\})}$$

$$\leq 2\limsup_{n\to\infty}\|T_k(\varrho_n)\|_{L^{\gamma+1}(\{\overline{T_k(\varrho)}\leq z_B\})}$$

$$\leq 2\left(\limsup_{n\to\infty}\|T_k(\varrho_n) - T_k(\varrho)\|_{L^{\gamma+1}((0,T)\times\Omega)} + \|T_k(\varrho) - \overline{T_k(\varrho)}\|_{L^{\gamma+1}((0,T)\times\Omega)} \right.$$

$$\left. + \|\overline{T_k(\varrho)}\|_{L^{\gamma+1}(\{\overline{T_k(\varrho)}\leq z_B\})} \right)$$

$$\leq 4\,\mathbf{osc}[\varrho_n \to \varrho]_{\gamma+1}((0,T)\times\Omega) + 2z_B(T|\Omega|)^{1/(\gamma+1)}. \tag{6.37}$$

The relations (6.35)–(6.37) imply that the right-hand side of (6.32) tends to zero with growing k, which completes the proof.

$\square$

6.6   Propagation of oscillations

Having collected all the necessary technical tools we focus on the central issue discussed at the beginning of this chapter, namely, *propagation of oscillations* in a sequence $\{\varrho_n\}_{n=1}^\infty$ of (non-negative) densities.

For simplicity of exposition, we shall assume that $\Omega \subset R^N$ is a bounded domain and prescribe the no-slip boundary conditions (1.41) for the velocity fields $\mathbf{u}_n$. Moreover, let the following hypotheses be satisfied:

1. The functions ϱ_n, $\mathbf{u}_n$ solve the continuity equation (1.34) in the sense of renormalized solutions in $\mathcal{D}'((0,T) \times R^N)$ provided they were extended to be zero outside Ω (see Definition 4.1); the functions ϱ_n are non-negative, and

$$\varrho_n \to \varrho \text{ weakly(-*) in } L^\infty(0,T;L^\gamma(\Omega)),$$

$$\mathbf{u}_n \to \mathbf{u} \text{ weakly in } L^2(0,T;W_0^{1,2}(\Omega;R^N)).$$

2. The functions ϱ_n, $\mathbf{u}_n$, ϑ_n satisfy the momentum equation (1.35) in the sense of distributions on $(0,T) \times \Omega$, with the stress tensor $\mathbb{S}_n$ determined through the constitutive law (1.37).

3. The pressure p is given by the constitutive equation (1.39), where the elastic component $p_e = p_e(\varrho)$ and the thermal component $\vartheta p_\vartheta(\varrho)$ satisfy hypotheses (3.1), (3.2) with

$$\gamma > \frac{N}{2}.$$

4. The temperature $\vartheta_n \geq 0$, which appears in the thermal pressure, satisfies

$$\vartheta_n \to \vartheta \text{ weakly in } L^2(0,T;W^{1,2}(\Omega)). \tag{6.38}$$

5. The total kinetic energy

$$\int_\Omega \varrho_n |\mathbf{u}_n|^2 \text{ is bounded in } L^\infty(0,T) \quad \text{for } n = 1,2,\ldots,$$

and

$$\mathbf{f}_n \to \mathbf{f} \text{ weakly(-*) in } L^\infty((0,T) \times \Omega).$$

To begin with, it is easy to observe that, by virtue of Lemma 4.2, the limit functions ϱ, $\mathbf{u}$ solve the continuity equation (1.35) in $\mathcal{D}'((0,T) \times R^N)$ provided ϱ, $\mathbf{u}$ were extended to be zero outside Ω. Moreover, the local pressure estimates stated in Proposition 5.1 yield

$$p_e(\varrho_n) \text{ bounded in } L^{(\gamma+\omega)/\gamma}(O) \quad \text{for a certain } \omega > 0$$

on any compact $O \subset ((0,T) \times \Omega)$.

Consequently, we are allowed to apply Proposition 6.2 to deduce a bound on the oscillations defect measure:

$$\mathbf{osc}_{\gamma+1}[\varrho_n \to \varrho](O) \leq c(|O|)$$

for any compact $O \subset ((0,T) \times \Omega)$. Moreover, as the value of the oscillations defect measure depends only on the measure of O, the set $(0,T) \times \Omega$ is bounded, and (by definition) $\varrho_n \equiv \varrho \equiv 0$ outside Ω, we infer

$$\mathbf{osc}_{\gamma+1}[\varrho_n \, , \, \varrho]((0,T) \times R^N) < \infty.$$

In accordance with Propsition 6.3, this means that ϱ, $\mathbf{u}$ solve (1.34) in the sense of renormalized solutions in $\mathcal{D}'((0,T) \times R^N)$ (see Definition 4.1).

Pursuing the general ideas delineated in the introductory part of this chapter (see (6.1)), the amplitude of oscillations in the sequence $\{\varrho_n\}_{n=1}^{\infty}$ will be measured by a quantity $\mathbf{dft}[\varrho_n \to \varrho]$,

$$\mathbf{dft}[\varrho_n \to \varrho](t) \equiv \int_{\Omega} \Big(\overline{\varrho \log(\varrho)} - \varrho \log(\varrho)\Big)(t, \mathbf{x})\, \mathrm{d}\mathbf{x}, \quad t \in [0,T], \quad (6.39)$$

where

$$\overline{\varrho \log(\varrho)} \text{ is a weak limit of the sequence } \{\varrho_n \log(\varrho_n)\}_{n=1}^{\infty}$$

in the Lebesgue spaces $L^{\omega}((0,T) \times \Omega)$ for a certain $1 \le \omega < \gamma$ (cf. Proposition 2.1).

As ϱ_n is a renormalized solution of the continuity equation, one is tempted to take $B(\varrho) = \varrho \log(\varrho)$ in (4.13). However, this is not possible in general as the corresponding function $b(\varrho) = \varrho$ is linear in ϱ, and we would need ϱ_n to be square integrable to define properly the term $\varrho_n \operatorname{div}_x \mathbf{u}_n$.

In order to avoid this rather technical difficulty, we introduce auxilliary functions

$$L_k(\varrho) \equiv \varrho \int_1^{\varrho} \frac{T_k(z)}{z^2}\, \mathrm{d}z, \qquad (6.40)$$

where T_k are the cut-off functions defined by (2.11).

Now, the equation

$$\partial_t L_k(\varrho_n) + \operatorname{div}_x(L_k(\varrho_n)\mathbf{u}_n) + T_k(\varrho_n)\operatorname{div}_x \mathbf{u}_n = 0 \qquad (6.41)$$

holds in $\mathcal{D}'((0,T) \times R^N)$, and letting $n \to \infty$ we obtain

$$\partial_t \overline{L_k(\varrho)} + \operatorname{div}_x(\overline{L_k(\varrho)}\mathbf{u}) + \overline{T_k(\varrho)\operatorname{div}_x \mathbf{u}} = 0, \qquad (6.42)$$

where

$$L_k(\varrho_n) \to \overline{L_k(\varrho)} \text{ in } C([0,T]; L^{\gamma}_{\text{weak}}(\Omega)),$$

$$T_k(\varrho_n)\operatorname{div}_x \mathbf{u}_n \to \overline{T_k(\varrho)\operatorname{div}_x \mathbf{u}} \text{ weakly in } L^2((0,T) \times \Omega).$$

In particular, since all functions $\overline{L_k(\varrho)}$ are continuous on $[0,T]$ with respect to the weak topology on $L^1(\Omega)$, and

$$L_k(\varrho) = \varrho \log(\varrho) \text{ for } 0 \le \varrho \le k, \quad 0 \le L_k(\varrho) \le \varrho \log(\varrho) \text{ otherwise}, \qquad (6.43)$$

the function $\overline{\varrho \log(\varrho)}$ belongs to the space $C([0,T]; L^\omega_{\text{weak}}(\Omega))$ for any $1 \leq \omega < \gamma$. By virtue of Proposition 4.3, $\varrho \log(\varrho)$ is a strongly continuous function of $t \in [0,T]$ with values in $L^1(\Omega)$; whence

$$t \mapsto \mathbf{dft}[\varrho_n \to \varrho](t) \text{ is continuous} \quad \text{for } t \in [0,T]. \tag{6.44}$$

Being a renormalized solution of (1.34), the functions ϱ, $\mathbf{u}$ satisfy

$$\partial_t L_k(\varrho) + \text{div}_x(L_k(\varrho)\mathbf{u}) + T_k(\varrho)\text{div}_x\,\mathbf{u} = 0 \quad \text{in } \mathcal{D}'((0,T) \times R^N);$$

therefore, by virtue of (6.42),

$$\partial_t\big(\overline{L_k(\varrho)} - L_k(\varrho)\big) + \text{div}_x\big((\overline{L_k(\varrho)} - L_k(\varrho))\mathbf{u}\big) + \overline{T_k(\varrho)\text{div}_x\,\mathbf{u}} - \overline{T_k(\varrho)}\text{div}_x\,\mathbf{u}$$
$$= (T_k(\varrho) - \overline{T_k(\varrho)})\text{div}_x\,\mathbf{u} \text{ in } \mathcal{D}'((0,T) \times R^N). \tag{6.45}$$

Thus taking a test function $\varphi(t,\mathbf{x}) = \psi(t)\eta(\mathbf{x})$, with $\psi \in \mathcal{D}(0,T)$, and $\eta \in \mathcal{D}(R^N)$, $\eta \equiv 1$ on an open neighbourhood of $\overline{\Omega}$, we deduce from (6.45) that

$$\int_0^T \int_\Omega \big(L_k(\varrho) - \overline{L_k(\varrho)}\big)\partial_t\psi\,d\mathbf{x}\,dt + \int_0^T \int_\Omega \big(\overline{T_k(\varrho)\text{div}_x\,\mathbf{u}} - \overline{T_k(\varrho)}\text{div}_x\,\mathbf{u}\big)\psi\,d\mathbf{x}\,dt$$
$$= \int_0^T \int_\Omega \big(T_k(\varrho) - \overline{T_k(\varrho)}\big)\text{div}_x\,\mathbf{u}\,\psi\,d\mathbf{x}\,dt,$$

which yields

$$\int_\Omega (\overline{L_k(\varrho)} - L_k(\varrho))(\tau_2)\,d\mathbf{x} - \int_\Omega (\overline{L_k(\varrho)} - L_k(\varrho))(\tau_1)\,d\mathbf{x}$$
$$+ \int_{\tau_1}^{\tau_2} \int_\Omega \overline{T_k(\varrho)\text{div}_x\,\mathbf{u}} - \overline{T_k(\varrho)}\text{div}_x\,\mathbf{u}\,d\mathbf{x}\,dt$$
$$= \int_{\tau_1}^{\tau_2} \int_\Omega \big(T_k(\varrho) - \overline{T_k(\varrho)}\big)\text{div}_x\,\mathbf{u}\,d\mathbf{x}\,dt \quad \text{for any } \tau_1 \leq \tau_2, \ \tau_1,\tau_2 \in [0,T]. \tag{6.46}$$

Now, similarly as in the proof of Proposition 6.3, we can estimate

$$\left| \int_{\tau_1}^{\tau_2} \int_\Omega \big(T_k(\varrho) - \overline{T_k(\varrho)}\big)\text{div}_x\,\mathbf{u}\,d\mathbf{x}\,dt \right|$$
$$\leq \|\text{div}_x\,\mathbf{u}\|_{L^2((0,T)\times\Omega)}\|\overline{T_k(\varrho)} - T_k(\varrho)\|_{L^2((0,T)\times\Omega)},$$

where

$$\|\overline{T_k(\varrho)} - T_k(\varrho)\|_{L^2((0,T)\times\Omega)} \leq \|\overline{T_k(\varrho)} - T_k(\varrho)\|^\omega_{L^1((0,T)\times\Omega)}$$
$$\times \|\overline{T_k(\varrho)} - T_k(\varrho)\|^{1-\omega}_{L^{\gamma+1}((0,T)\times\Omega)}$$

for a certain $\omega \in (0,1)$. Seeing that

$$\|\overline{T_k(\varrho)} - T_k(\varrho)\|_{L^1((0,T)\times\Omega)} \to 0 \quad \text{for } k \to \infty,$$

and

$$\|\overline{T_k(\varrho)} - T_k(\varrho)\|_{L^{\gamma+1}((0,T)\times\Omega)} \leq \mathbf{osc}[\varrho_n \to \varrho]_{\gamma+1}((0,T) \times R^N),$$

we conclude that

$$\left|\int_{\tau_1}^{\tau_2} \int_\Omega \left(T_k(\varrho) - \overline{T_k(\varrho)}\right)\mathrm{div}_x\,\mathbf{u}\,\mathrm{d}x\,\mathrm{d}t\right| \to 0 \quad \text{for } k \to \infty.$$

Consequently, one can let $k \to \infty$ in (6.46) to obtain

$$\mathbf{dft}[\varrho_n \to \varrho](\tau_2) - \mathbf{dft}[\varrho_n \to \varrho](\tau_1) = \lim_{k\to\infty} \int_{\tau_1}^{\tau_2} \int_\Omega \overline{T_k(\varrho)\mathrm{div}_x\,\mathbf{u}} - \overline{T_k(\varrho)}\mathrm{div}_x\,\mathbf{u}\,\mathrm{d}x\,\mathrm{d}t$$

(6.47)

for any $\tau_1 \leq \tau_2$.

As for the expression on the right-hand side of (6.47), Proposition 6.1 yields

$$\int_{\tau_1}^{\tau_2} \int_K \overline{T_k(\varrho)\,\mathrm{div}_x\,\mathbf{u}} - \overline{T_k(\varrho)}\,\mathrm{div}_x\,\mathbf{u}\,\mathrm{d}x\,\mathrm{d}t$$

$$= \lim_{n\to\infty} \int_{\tau_1}^{\tau_2} \int_K \overline{T_k(\varrho)}\mathrm{div}_x\,\mathbf{u} - T_k(\varrho_n)\mathrm{div}_x\,\mathbf{u}_n\,\mathrm{d}x\,\mathrm{d}t$$

$$= \frac{1}{2\mu+\lambda} \int_{\tau_1}^{\tau_2} \int_K \left(\overline{p_e(\varrho)} + \vartheta\,\overline{p_\vartheta(\varrho)}\right)\overline{T_k(\varrho)}$$

$$- \frac{1}{2\mu+\lambda} \lim_{n\to\infty} \int_{\tau_1}^{\tau_2} \int_K \left(p_e(\varrho_n) + \vartheta_n p_\vartheta(\varrho_n)\right)T_k(\varrho_n)\mathrm{d}x\,\mathrm{d}t \qquad (6.48)$$

for any $\tau_1, \tau_2 \in (0,T)$, and any compact $K \subset \Omega$. Note that, by virtue of hypotheses (3.1), (3.2) and the fact that ϱ_n is a renormalized solution of (1.34),

$$p_\vartheta(\varrho_n) \to \overline{p_\vartheta(\varrho)} \text{ in } C([0,T]; L^N_{\text{weak}}(\Omega)),$$

which, together with (6.38), yields

$$\overline{\vartheta\,p_\vartheta(\varrho)} = \vartheta\,\overline{p_\vartheta(\varrho)}.$$

The relations (6.47), (6.48) express an interesting link between the continuity and momentum equations. As we shall see, the density oscillations are damped in the region where the pressure is a non-decreasing function of the density, and they are enhanced if this is not the case. To demonstrate this, we distinguish three different situations analyzed in the remaining part of this section.

6.6.1 Monotone pressure

We start with the most natural situation when both the elastic pressure component p_e and p_ϑ are *non-decreasing functions* of the density ϱ.

As we have already seen in (6.28),

$$\lim_{n\to\infty} \int_O p_e(\varrho_n)T_k(\varrho_n) - \overline{p_e(\varrho)}\ \overline{T_k(\varrho)}\,\mathrm{d}\mathbf{x}\,\mathrm{d}t \geq 0 \qquad (6.49)$$

for any compact $O \subset ((0,T) \times \Omega)$. Moreover, by the same token,

$$\overline{p_\vartheta(\varrho)\ T_k(\varrho)} \geq \overline{p_\vartheta(\varrho)}\ \overline{T_k(\varrho)}. \qquad (6.50)$$

On the other hand, since ϱ_n is a renormalized solution of (1.34), we have

$$p_\vartheta(\varrho_n)T_k(\varrho_n) \to \overline{p_\vartheta(\varrho)T_k(\varrho)} \text{ in } C([0,T]; L^N_{\text{weak}}(\Omega)),$$

which, together with (6.38), yields

$$\vartheta_n p_\vartheta(\varrho_n)T_k(\varrho_n) \to \vartheta\ \overline{p_\vartheta(\varrho)\ T_k(\varrho)} \text{ weakly in } L^1((0,T) \times \Omega). \qquad (6.51)$$

We thereby obtain

$$\lim_{n\to\infty} \int_O \vartheta_n\ p_\vartheta(\varrho_n)\ T_k(\varrho_n)\mathrm{d}\mathbf{x}\,\mathrm{d}t \geq \int_O \vartheta\overline{p_\vartheta(\varrho)}\ \overline{T_k(\varrho)}\,\mathrm{d}\mathbf{x}\,\mathrm{d}t \qquad (6.52)$$

for any compact $O \subset (0,T) \times \Omega$.

By virtue of (6.49) and (6.52), the right-hand side of (6.47) is non-positive and, consequently,

$$\mathbf{dft}[\varrho_n \to \varrho](\tau_2) \leq \mathbf{dft}[\varrho_n \to \varrho](\tau_1) \text{ for any } \tau_2 \geq \tau_1. \qquad (6.53)$$

In other words we have arrived at the following conclusion:

> *The amplitude of density oscillations is a non-increasing function of time provided the pressure is a non-decreasing function of the density.*

The oscillations cannot be created unless they were present in the initial data.

Remark Strictly speaking, Theorem 2.11, together with the fact that $\mathbf{dft}[\varrho_n \to \varrho]$ vanish on $[0,T]$, implies strong convergence of the sequence $\{\varrho_n\}_{n=1}^\infty$ only on the set where the weak limit ϱ is positive. On the other hand, however, since ϱ_n are non-negative, they automatically converge to zero almost anywhere on the set where ϱ vanishes.

6.6.2 Convex pressure

As we have just discovered, monotone pressure prevents creation of density oscillations. If, in addition, the pressure p_e is a convex function of ϱ, the oscillations decay at a uniform rate with growing time. Similarly as in Proposition 6.2, we assume that the pressure p is given through a constitutive equation

$$p(\varrho, \vartheta) = p_{\mathrm{c}}(\varrho) \mid p_{\mathrm{m}}(\varrho) + \vartheta p_{\vartheta}(\varrho),$$

where

$$p_{\mathrm{c}} : [0, \infty) \to [0, \infty) \text{ is a convex function, } p_{\mathrm{c}}(0) = 0, \text{ and } p_{\mathrm{c}}(\varrho) \geq a\varrho^{\gamma}$$

for a certain $a > 0$,

$$p_{\mathrm{m}} : [0, \infty) \to [0, \infty) \text{ is non-decreasing, } p_{\mathrm{m}}(0) = 0,$$

and p_{ϑ} complies with hypothesis (3.2).

Utilizing inequality (6.26) together with (6.48) and (6.49), (6.52), we get

$$\int_{\tau_1}^{\tau_2} \int_K \overline{T_k(\varrho) \mathrm{div}_x \, \mathbf{u}} - \overline{T_k(\varrho)} \mathrm{div}_x \, \mathbf{u} \, \mathrm{dx} \, \mathrm{dt}$$

$$\leq -\frac{a}{2\mu + \lambda} \limsup_{n \to \infty} \int_{\tau_1}^{\tau_2} \int_K |T_k(\varrho_n) - T_k(\varrho)|^{\gamma+1} \, \mathrm{dx} \, \mathrm{dt} \qquad (6.54)$$

for any compact $K \subset \Omega$. As Ω is bounded, the same inequality holds also with K replaced by Ω.

Next, the informal idea is to express the right-hand side of the above inequality in terms of $\mathbf{dft}[\varrho_n \to \varrho]$. To this end, we exploit Hölder's inequality (2.1) to deduce

$$\int_{\tau_1}^{\tau_2} \int_\Omega |T_k(\varrho_n) - T_k(\varrho)|^{\gamma+1} \, \mathrm{dx} \, \mathrm{dt}$$

$$\geq c(\omega, |\Omega|) \int_{\tau_1}^{\tau_2} \|T_k(\varrho_n) - T_k(\varrho)\|_{L^\omega(\Omega)}^{\gamma+1} \, \mathrm{dt}$$

$$\geq c(\omega, |\Omega|) \int_{\tau_1}^{\tau_2} \|\varrho_n - \varrho\|_{L^\omega(\Omega)}^{\gamma+1} \, \mathrm{dt} - 2^{\gamma+1} c(\omega, |\Omega|) \int_{\tau_1}^{\tau_2} \|\varrho_n\|_{L^\omega(\{\varrho_n \geq k\})}^{\gamma+1} \, \mathrm{dt}$$

$$\geq c(\omega, |\Omega|) \int_{\tau_1}^{\tau_2} \|\varrho_n - \varrho\|_{L^\omega(\Omega)}^{\gamma+1} \, \mathrm{dt} - 2^{\gamma+1}(\tau_2 - \tau_1) c(\omega, |\Omega|) k^{(\omega-\gamma)(\gamma+1)/\omega}$$

$$\times \operatorname*{ess\,sup}_{t \in (0,T)} \|\varrho_n(t)\|_{L^\gamma(\Omega)}^{(\gamma+1)\gamma/\omega} \qquad (6.55)$$

for any $1 \leq \omega < \gamma$.

Combining (6.54), (6.55) with (6.47) we can let $k \to \infty$ to conclude that there exists a positive constant $c = c(\omega, |\Omega|)$ such that

$$\mathbf{dft}[\varrho_n \to \varrho](\tau_2) - \mathbf{dft}[\varrho_n \to \varrho](\tau_1) + c(\omega, |\Omega|) \limsup_{n \to \infty} \int_{\tau_1}^{\tau_2} \|\varrho_n(t) - \varrho(t)\|_{L^\omega(\Omega)}^{\gamma+1} \, \mathrm{dt} \leq 0$$

$$(6.56)$$

for any $1 \leq \omega < \gamma$.

In order to continue, we need the following assertion which can be verified by means of a direct computation: for any $\omega \in (1, \gamma)$ there exists $c(\omega) > 0$ such that

$$z \log(z) - y \log(y) \leq (1 + \log^+(y))(z - y) + c(\omega)(|z - y|^{1/2} + |z - y|^\omega) \quad (6.57)$$

for all $y, z \geq 0$, where $\log^+ \equiv \max\{\log, 0\}$ is the positive part of log.

Accordingly, we have

$$\int_\Omega \varrho_n \log(\varrho_n) - \varrho \log(\varrho) \, dx - \int_\Omega (1 + \log^+(\varrho))(\varrho_n - \varrho) \, dx$$
$$\leq c(\omega)(|\Omega|^{(2\omega-1)/2\omega} \|\varrho_n - \varrho\|^{1/2}_{L^\omega(\Omega)} + \|\varrho_n - \varrho\|^\omega_{L^\omega(\Omega)}).$$

Now, we introduce a function

$$A(z) \equiv c(\omega)(|\Omega|^{(2\omega-1)/2\omega} z^{1/2} + z^\omega) \quad \text{for } z \geq 0,$$

and

$$\Phi(z) \equiv c(\omega, |\Omega|)(A^{-1}(z))^{\gamma+1}, \quad (6.58)$$

where $c(\omega, |\Omega|)$ is the constant from (6.56).

Consequently,

$$\Phi\left(\int_\Omega \varrho_n \log(\varrho_n) - \varrho \log(\varrho) \, dx - \int_\Omega (1 + \log^+(\varrho))(\varrho_n - \varrho) \, dx\right)$$
$$\leq c(\alpha, |\Omega|)\|\varrho_n - \varrho\|^{\gamma+1}_{L^\alpha(\Omega)},$$

and we can let $n \to \infty$ in (6.56) to deduce

$$\mathbf{dft}[\varrho_n \to \varrho](\tau_2) - \mathbf{dft}[\varrho_n \to \varrho](\tau_1) + \int_{\tau_1}^{\tau_2} \Phi(\mathbf{dft}[\varrho_n \to \varrho](t)) \, dt \leq 0 \quad (6.59)$$

for all $\tau_1 \leq \tau_2$.

Inequality (6.59) implies *uniform decay of oscillations* with respect to time. More precisely, if $\tau_1 \leq \tau_2$, then

$$\mathbf{dft}[\varrho_n \to \varrho](\tau_2) \leq \chi(\tau_2 - \tau_1), \quad (6.60)$$

where χ is the unique solution of the initial-value problem

$$\chi'(t) + \Phi(\chi(t)) = 0, \quad \chi(0) = \mathbf{dft}[\varrho_n \to \varrho](\tau_1).$$

Possible oscillations in a density sequence $\{\varrho_n\}_{n=1}^\infty$ decay uniformly in time provided the elastic pressure component p_e is a strictly convex function of the density.

6.6.3 *General (non-monotone) (pdt) state equations*

Non-monotone pressure density dependence is related to the effect of binding molecular forces acting on a relatively long distance when the density is large enough. As an example of this phenomenon, consider a pressure law in the form

$$p(\varrho, \vartheta) = p_e(\varrho) + \vartheta p_\vartheta(\varrho),$$

where (i) p'_e is bounded from below, that is,

$$p'_e(\varrho) \geq -\underline{p} \quad \text{for all } \varrho > 0;$$

(ii) p_e is a strictly increasing function of ϱ provided ϱ is large enough, more precisely, there exists $\varrho_0 \geq 0$ and $\delta > 0$ such that

$$p'_e(\varrho) \geq \delta > 0 \quad \text{for all } \varrho \geq \varrho_0;$$

and the thermal pressure component p_ϑ satisfies hypothesis (3.2).

Under the above hypotheses, it is easy to see that the elastic pressure p_e can be written as

$$p_e(\varrho) = p_m(\varrho) - r(\varrho) \tag{6.61}$$

with p_m a non-decreasing function of ϱ, and

$$r \in C^2[0, \infty), \quad r \geq 0, \quad r(\varrho) \equiv 0 \text{ for } \varrho \geq \varrho_r \tag{6.62}$$

for a certain $\varrho_r \geq 0$.

Using (6.49), (6.52) we deduce from (6.47) that

$$\mathbf{dft}[\varrho_n \to \varrho](\tau_2) - \mathbf{dft}[\varrho_n \to \varrho](\tau_1)$$

$$\leq \frac{1}{2\mu + \lambda} \lim_{k \to \infty} \left(\lim_{n \to \infty} \int_{\tau_1}^{\tau_2} \int_\Omega r(\varrho_n) T_k(\varrho_n) - \overline{r(\varrho)} \; \overline{T_k(\varrho)} \, dx \, dt \right). \tag{6.63}$$

As the sequence $\{\varrho_n\}_{n=1}^\infty$ is bounded in $L^\infty(0, T; L^\gamma(\Omega))$, and r is a bounded function, we have

$$\lim_{k \to \infty} \left(\lim_{n \to \infty} \int_{\tau_1}^{\tau_2} \int_\Omega r(\varrho_n) T_k(\varrho_n) - \overline{r(\varrho)} \; \overline{T_k(\varrho)} \, dx \, dt \right)$$

$$= \lim_{n \to \infty} \int_{\tau_1}^{\tau_2} \int_\Omega r(\varrho_n)\varrho_n - \overline{r(\varrho)} \; \varrho \, dx \, dt. \tag{6.64}$$

Since the function r is twice continuously differentiable and compactly supported on $[0, \infty)$, there exists $\Lambda > 0$ such that both $\varrho \mapsto \Lambda \varrho \log(\varrho) - \varrho r(\varrho)$ and $\varrho \mapsto \Lambda \varrho \log(\varrho) + r(\varrho)$ are convex functions of ϱ.

As a consequence of weak lower semi-continuity of convex functionals (see Corollary 2.2), we obtain

$$\lim_{n\to\infty} \int_{\tau_1}^{\tau_2} \int_\Omega r(\varrho_n)\varrho_n - \overline{r(\varrho)}\varrho \, d\mathbf{x} \, dt$$

$$\leq \Lambda \int_{\tau_1}^{\tau_2} \int_\Omega \overline{\varrho \log(\varrho)} - \varrho \log(\varrho) \, d\mathbf{x} \, dt + \int_{\tau_1}^{\tau_2} \int_\Omega (r(\varrho) - \overline{r(\varrho)})\varrho \, d\mathbf{x} \, dt. \quad (6.65)$$

Furthermore r is non-negative and, consequently,

$$\int_{\tau_1}^{\tau_2} \int_\Omega (r(\varrho) - \overline{r(\varrho)})\varrho \, d\mathbf{x} \, dt \leq \int_{\{\varrho \leq \varrho_r\}} (r(\varrho) - \overline{r(\varrho)})\varrho \, d\mathbf{x} \, dt$$

$$\leq \Lambda \int_{\{\varrho \leq \varrho_r\}} (\overline{\varrho \log(\varrho)} - \varrho \log(\varrho))\varrho \, d\mathbf{x} \, dt$$

$$\leq \Lambda \varrho_r \int_{\tau_1}^{\tau_2} \int_\Omega \overline{\varrho \log(\varrho)} - \varrho \log(\varrho) \, d\mathbf{x}. \quad (6.66)$$

Relations (6.63)–(6.66) yield an inequality

$$\mathbf{dft}[\varrho_n \to \varrho](\tau_2) \leq \mathbf{dft}[\varrho_n \to \varrho](\tau_1) + \omega \int_{\tau_1}^{\tau_2} \mathbf{dft}[\varrho_n \to \varrho](t) \, dt$$

with a certain constant $\omega \geq 0$. Applying Gronwall's lemma we infer

$$\mathbf{dft}[\varrho_n \to \varrho](\tau_2) \leq \mathbf{dft}[\varrho_n \to \varrho](\tau_1) \exp(\omega(\tau_2 - \tau_1)) \quad (6.67)$$

for any $\tau_1 \leq \tau_2$.

A general (non-monotone) pressure density relation can enhance density oscillations but these still cannot be created in a finite time.

6.7 Weak stability revisited

6.7.1 Momentum equation

In Chapter 4, we examined the problem of *weak sequential stability* of the set of variational solutions of the Navier–Stokes system (1.34)–(1.36). More precisely, it was shown that any sequence of quantities ϱ_n, $\mathbf{u}_n$, p_n, and $\mathbb{S}_n$ solving (1.34) and (1.35) in the sense of distributions admits a weak limit which is a distributional solution of the same problem. The thermal energy equation (1.36) was left apart as it contains certain nonlinear terms whose "*weak continuity*" is more delicate to prove.

Pursuing further this programme we exploit the main achievement of the preceding section, namely, strong convergence of the densities in $L^1((0,T) \times \Omega)$. Indeed we have seen that, under very mild assumptions on the constitutive equation for the pressure, the oscillations in a sequence $\{\varrho_n\}_{n=1}^{\infty}$ cannot occur in a finite time unless they were already present in the initial data. Accordingly, under the hypotheses made at the beginning of Section 6.6, we have

$$\varrho_n \to \varrho \ \text{(strongly) in } L^\omega((0,T) \times \Omega) \quad \text{for any } 1 \le \omega < \gamma \qquad (6.68)$$

provided

$$\varrho_n(0) \to \varrho(0) \text{ in } L^1(\Omega).$$

Now we assert that in fact

$$\varrho_n \to \varrho \text{ in } C([0,T]; L^\omega(\Omega)) \quad \text{for any } 1 \le \omega < \gamma. \qquad (6.69)$$

To see this, it is enough to show that

$$T_k(\varrho_n) \to T_k(\varrho) \text{ in } C([0,T], L^1(\Omega)) \quad \text{for any fixed } k \ge 1, \qquad (6.70)$$

where T_k are the cut-off functions defined in (2.11).

In order to prove (6.70), we first show an auxilliary result:

Lemma 6.2 *Let $\Omega \subset R^N$ be a bounded measurable set. Furthermore, let $\{v_n\}_{n=1}^{\infty}$ be a sequence of functions such that*

$$v_n \to v \text{ in } C([0,T]; L_{\text{weak}}^2(\Omega)),$$

and

$$|v_n|^2 \to |v|^2 \text{ in } C([0,T]; L_{\text{weak}}^1(\Omega)).$$

Then

$$v_n \to v \text{ in } C([0,T]; L^2(\Omega)).$$

Proof   Since $L^2(\Omega)$ is a Hilbert space, it follows immediately that

$$v_n(t) \to v(t) \text{ in } L^2(\Omega) \quad \text{for any } t \in [0,T].$$

Moreover, the norm of the difference $v_n - v$ can be written as

$$\|v_n(t) - v(t)\|_{L^2(\Omega)}^2 = \int_\Omega |v_n(t)|^2 - |v(t)|^2 \, d\mathbf{x} + 2 \int_\Omega (v - v_n)(t)v(t) \, d\mathbf{x},$$

where the first integral on the right-hand side tends to zero for $n \to \infty$ uniformly in t.

Furthermore, since the trajectory $\cup_{t \in [0,T]} v(t)$ is a compact set in $L^2(\Omega)$, there exists for any $\varepsilon > 0$ a finite number of times $t_1, \ldots, t_{n(\varepsilon)}$ such that for any $t \in [0, T]$ there exists t_i such that

$$\|v(t_i) - v(t)\|_{L^2(\Omega)} < \varepsilon.$$

Consequently,

$$\left| \int_\Omega (v - v_n)(t)v(t) \, d\mathbf{x} \right| \leq \left| \int_\Omega (v - v_n)(t)v(t_i) \, d\mathbf{x} \right| + \varepsilon \sup_{t \in [0,T]} \|(v - v_n)(t)\|_{L^2(\Omega)};$$

and we conclude

$$\int_\Omega (v - v_n)(t)v(t) \, d\mathbf{x} \to 0 \quad \text{uniformly with respect to } t \in [0, T]$$

which completes the proof.

$\square$

Since both ϱ_n and ϱ are renormalized solutions of (1.34), we can apply Lemma 6.2 to $v_n = T_k(\varrho_n)$, $v = T_k(\varrho)$ to obtain (6.68) and, consequently, (6.69).

With the local pressure estimates stated in Proposition 5.1 at hand, it is now easy to see that

$$p_e(\varrho_n) \to p_e(\varrho) \quad \text{in } \mathcal{D}'((0,T) \times \Omega), \tag{6.71}$$

and, by the same token,

$$\vartheta_n p_\vartheta(\varrho_n) \to \vartheta p_\vartheta(\varrho) \text{ weakly in } L^2((0,T) \times \Omega). \tag{6.72}$$

Finally,

$$\varrho_n \mathbf{f}_n \to \varrho \mathbf{f} \text{ weakly(-*) in } L^\infty(0,T; L^\gamma(\Omega)). \tag{6.73}$$

Thus we have established *weak sequential compactness* of the set of bounded energy variational solutions to the system (1.34), (1.35), where the pressure p and the stress tensor $\mathbb{S}$ are determined through the constitutive equations (1.39) and (1.37), respectively.

Theorem 6.1 *Let $\Omega \subset R^N$, $N \geq 2$, be a bounded domain of class $C^{2+\nu}$, $\nu > 0$. Furthermore, let ϱ_n, $\mathbf{u}_n$ (and ϑ_n) be variational solutions of the equation of continuity (1.34) and the momentum equation (1.35) on $(0,T) \times \Omega$ in the sense of Definitions 4.2, 4.3; with the pressure p_n and the viscous stress $\mathbb{S}_n$ determined through the constitutive equations (1.39) and (1.37), which comply with hypotheses (3.1), (3.2) and (1.38) respectively with*

$$\gamma > \frac{N}{2}.$$

Finally, assume that the total mechanical energy $E_{\mathrm{mech}}[\varrho_n(t), (\varrho_n \mathbf{u}_n)(t)]$ defined through (4.36) is bounded for any $t \in [0,T]$ by a a constant E_0 independent of $n = 1, 2, \ldots$; and

$$\varrho_n \to \varrho \ \text{weakly(-*) in } L^\infty(0,T; L^\gamma(\Omega)),$$

$$\varrho_n(0+) \to \varrho(0+) \ \text{in } L^1(\Omega),$$

$$\mathbf{u}_n \to \mathbf{u} \ \text{weakly in } L^2(0,T; W_0^{1,2}(\Omega; R^N)),$$

$$\vartheta_n \geq 0, \vartheta_n \to \overline{\vartheta} \ \text{weakly in } L^2(0,T; W^{1,2}(\Omega)),$$

$$\mathbf{f}_n \to \mathbf{f} \ \text{weakly(-*) in } L^\infty((0,T) \times \Omega).$$

Then the limit functions ϱ, $\mathbf{u}$, $\overline{\vartheta}$ are variational solutions of the system (1.34), (1.35) in the sense of Definitions 4.2, 4.3, with

$$p(\varrho, \overline{\vartheta}) = p_{\mathrm{e}}(\varrho) + \overline{\vartheta} p_\vartheta(\varrho),$$

$$\mathbb{S} = \mu(\nabla_x \mathbf{u} + \nabla_x \mathbf{u}^t) + \lambda \mathbb{I} \, \mathrm{div}_x \, \mathbf{u},$$

and the external force density $\mathbf{f}$.

Remark In view of Proposition 4.3, the instantaneous values of the density $\varrho_n(t)$ are continuous in time, in particular, the value of $\varrho_n(0+)$ is correctly defined.

6.7.2 Thermal energy equation

In order to establish a similar result for the thermal energy equation (1.36), we have to verify first that the temperature ϑ_n solving (1.36) in the sense of Definition 4.5 is precompact in the L^1-topology at least on the set where the limit density ϱ is strictly positive. More specifically, we will show that

$$\varrho_n Q(\vartheta_n)^2 \to \varrho \overline{Q(\vartheta)}^2 \quad \text{in } \mathcal{D}'((0,T) \times \Omega). \tag{6.74}$$

In accordance with (4.51), we shall assume we have bounds on the sequence $\{Q(\vartheta_n)\}_{n=1}^\infty$ in the space $L^2(0,T; W^{1,2}(\Omega))$ and that

$$Q(\vartheta_n) \to \overline{Q(\vartheta)} \ \text{weakly in } L^2(0,T; W^{1,2}(\Omega)).$$

Consequently, in order to see (6.74), it is enough to prove

$$\varrho_n Q(\vartheta_n) \to \overline{\varrho Q(\vartheta)} \text{ (strongly) in } L^2(0,T;W^{-1,2}(\Omega)), \qquad (6.75)$$

where the dual space $W^{-1,2}$ was introduced in Section 2.1 of Chapter 2.

As

$$\varrho_n Q(\vartheta_n) \text{ are bounded in } L^2(0,T;L^{q_2}(\Omega)) \quad \text{with } q_2 = m_2 > \frac{2N}{N+2},$$

where m_2 is given in (3.47), the Lebesgue space $L^{q_2}(\Omega)$ is compactly imbedded into $W^{-1,2}(\Omega)$ (see Theorem 2.8). Thus compactness may fail only if there were oscillations of $\varrho_n Q(\vartheta_n)$ with respect to time. Note in this respect that the variational formulation introduced in Section 4.3 replaces *equation* (1.36) by a family of integral *inequalities* (4.41); in particular, we do not know if the instantaneous values of $\varrho_n Q(\vartheta_n)$ are weakly continuous in t as is the case for the momentum $\varrho_n \mathbf{u}_n$.

In order to show (6.75), we establish a variant of the celebrated *Aubin–Lions lemma*.

Lemma 6.3 *Let* $\{v_n\}_{n=1}^{\infty}$ *be a sequence of functions such that*

$$v_n \text{ are bounded in } L^2(0,T;L^q(\Omega)) \cap L^{\infty}(0,T;L^1(\Omega)), \quad \text{with } q > \frac{2N}{N+2}.$$

Furthermore, assume that

$$\partial_t v_n \geq g_n \quad \text{in } \mathcal{D}'((0,T) \times \Omega), \qquad (6.76)$$

where

$$g_n \text{ are bounded in } L^1(0,T;W^{-m,r}(\Omega))$$

for a certain $m \geq 1$, $r > 1$.
Then $\{v_n\}_{n=1}^{\infty}$ *contains a subsequence such that*

$$v_n \to v \text{ in } L^2(0,T;W^{-1,2}(\Omega)).$$

Proof First of all, we can assume that $W^{-1,2}(\Omega)$ is continuously imbedded in $W^{-m,r}(\Omega)$ taking m large as the case may be. Moreover, as the Lebesgue space $L^q(\Omega)$ is compactly imbedded into $W^{-1,2}(\Omega)$ (see Theorem 2.8), we have

$$\|v\|_{W^{-1,2}(\Omega)} \leq \omega\|v\|_{L^q(\Omega)} + c(\omega)\|v\|_{W^{-m,r}(\Omega)} \quad \text{for any } \omega > 0.$$

Consequently, it suffices to show that

$$v_n \to v \quad \text{in } L^2(0,T;W^{-m,r}(\Omega)). \qquad (6.77)$$

We may assume m is so large that $L^1(\Omega)$ is compactly imbedded into $W^{-m,r}(\Omega)$; therefore (6.77) follows as soon as we establish pointwise convergence of $\{v_n\}_{n=1}^{\infty}$, specifically,

$$v_n(t) \rightarrow v(t) \text{ weakly in } W^{-m,r}(\Omega) \quad \text{for a.a. } t \in [0,T]. \tag{6.78}$$

Let $\{\eta_\nu\}_{\nu=1}^{\infty}$,

$$\eta_\nu \geq 0, \quad \eta_\nu \in \mathcal{D}(\Omega),$$

be a family of functions such that $\mathrm{span}\{\eta_\nu\}_{\nu=1}^{\infty}$ is dense in $C_0(\overline{\Omega})$.

It follows from hypothesis (6.76) that the sequence of functions

$$t \mapsto \int_{\Omega} v_n(t)\eta_\nu \, \mathrm{d}\mathbf{x}, \quad n = 1, 2, \ldots$$

is bounded in the space $BV[0,T]$ of functions with bounded variation on $[0,T]$ for any fixed ν (see Chapter 5 in [42] for the definition and basic properties of BV-functions). Indeed it is a routine matter to deduce an inequality

$$\int_{\Omega} v_n(t_1)\eta_\nu \, \mathrm{d}\mathbf{x} \leq \int_{\Omega} v_n(t_2)\eta_\nu \mathrm{d}\mathbf{x} + \int_{t_1}^{t_2} \int_{\Omega} g_n \eta_\nu \, \mathrm{d}\mathbf{x} \, \mathrm{d}t$$

for any Lebesgue points $t_1 \leq t_2$ of the function $t \mapsto \int_{\Omega} v_n(t)\eta_\nu \, \mathrm{d}\mathbf{x}$. Consequently, the functions

$$t \mapsto \int_{\Omega} v_n(t)\eta_\nu \, \mathrm{d}\mathbf{x} + \int_0^t \int_{\Omega} g_n \eta_\nu \, \mathrm{d}\mathbf{x} \, \mathrm{d}s$$

are essentially bounded and non-decreasing on $(0,T)$, which yields the desired result.

Accordingly, as $BV[0,T]$ is compactly imbedded into $L^1(0,T)$ (see Theorem 4 of Section 5.2.3 in [42]), we can assume

$$\int_{\Omega} v_n(t)\eta_\nu \, \mathrm{d}\mathbf{x} \rightarrow V[\eta_\nu](t) \quad \text{as } n \rightarrow \infty \quad \text{for a.a. } t \in [0,T],$$

where the mapping $\eta_\nu \mapsto V[\eta_\nu]$ is linear. As v_n are bounded in the space $L^\infty(0,T;L^1(\Omega))$ and $\mathrm{span}\{\eta_\nu\}_{\nu=1}^{\infty}$ is dense in $C_0(\overline{\Omega})$, there exists $v(t) \in C_0(\overline{\Omega})^*$ such that

$$\int_{\Omega} v_n(t)\eta \, \mathrm{d}\mathbf{x} \rightarrow \, <v(t), \eta> \quad \text{as } n \rightarrow \infty \quad \text{for a.a. } t \in (0,T) \text{ and any } \eta \in C^0(\overline{\Omega}).$$

Taking m large enough we may assume $W_0^{m,r'}(\Omega) \subset C_0(\overline{\Omega})$ and, consequently, (6.78) follows.

$\square$

Now as $\varrho_n Q(\vartheta_n)$ are supposed to satisfy (4.41), Lemma 6.3 can be applied to $v_n = \varrho_n Q(\vartheta_n)$ to obtain (6.75) therefore (6.74).

As a corollary of relations (6.69) and (6.74), we get

$$\int_0^T \int_\Omega \varrho Q(\vartheta_n)^2 \, \mathrm{d}\mathbf{x} \, \mathrm{d}t$$

$$= \int_0^T \int_\Omega (\varrho - \varrho_n) Q(\vartheta_n)^2 \, \mathrm{d}\mathbf{x} \, \mathrm{d}t + \int_0^T \int_\Omega \varrho_n Q(\vartheta_n)^2 \, \mathrm{d}\mathbf{x} \, \mathrm{d}t$$

$$\to \int_0^T \int_\Omega \varrho \overline{Q(\vartheta)}^2 \, \mathrm{d}\mathbf{x} \, \mathrm{d}t \quad \text{as } n \to \infty.$$

This can be interpreted as convergence of norms in a weighted Hilbert space L_ϱ^2, in particular, we have

$$Q(\vartheta_n)(t,\mathbf{x}) \to \overline{Q(\vartheta)}(t,\mathbf{x}) \quad \text{for a.a. } (t,\mathbf{x}) \in \{(t,\mathbf{x}) \in (0,T) \times \Omega \mid \varrho(t,\mathbf{x}) > 0\}.$$

Since Q is strictly increasing, it follows (in notation of Theorem 6.1) that

$$\vartheta_n \to \overline{\vartheta} \text{ (strongly) in } L^1(\{\varrho > 0\}). \tag{6.79}$$

Strong convergence of the temperature established in (6.79) is sufficient for the nonlinear terms in (4.41) to converge to their limit counterparts:

$$\varrho_n Q(\vartheta_n) \to \varrho Q(\overline{\vartheta}) \text{ weakly in } L^2(0,T;L^{q_2}(\Omega)),$$

$$\varrho_n Q(\vartheta_n)\mathbf{u}_n \to \varrho Q(\overline{\vartheta})\mathbf{u} \text{ weakly in } L^2(0,T;L^{c_2}(\Omega)),$$

where $q_2 = m_2$ and c_2 have the same values as in Theorem 3.1.

Furthermore, if p_ϑ obeys hypothesis (3.2), we have

$$\vartheta_n p_\vartheta(\varrho_n)\mathrm{div}_x \, \mathbf{u}_n \to \overline{\vartheta} p_\vartheta(\varrho)\mathrm{div}_x \, \mathbf{u} \text{ weakly in } L^1((0,T) \times \Omega). \tag{6.80}$$

We assert the following supplement of Theorem 6.1 concerning *weak sequential stability* (compactness) for the full system (1.34)–(1.36) with the boundary conditions (1.41), (1.42).

Theorem 6.2 *In addition to the hypotheses of Theorem 6.1, assume that ϱ_n, $\mathbf{u}_n$ and ϑ_n are renormalized solutions of the problem (4.38), (4.39) with*

$$\vartheta_n(0) \geq \vartheta_{0,n}(\geq 0)$$

in the sense of Definition 4.4, and such that

$$Q(\vartheta_n) \text{ are bounded in } L^2(0,T;W^{1,2}(\Omega)),$$

and with

$$\kappa(\vartheta_n) \text{ bounded in } L^1((0,T)\times\Omega),$$

where κ is determined through (4.43). Furthemore assume that the energy inequality

$$E_n(\tau) \leq E_n(0) + \int_0^\tau \int_\Omega \varrho_n \mathbf{f}_n \cdot \mathbf{u}_n \, \mathbf{dx}\, \mathrm{dt}$$

holds for a.a. $\tau \in [0,T]$ with

$$E_n(0) = \int_\Omega \frac{1}{2}\frac{|\varrho_n\mathbf{u}_n(0+)|^2}{\varrho_n(0+)} + \varrho_n(0+)P_\mathrm{e}(\varrho_n(0+)) + \varrho_n(0+)Q(\vartheta_{0,n}) \, \mathbf{dx},$$

where

$$\varrho_n(0+) \to \varrho_0 \quad \text{in } L^\gamma(\Omega),$$

$$(\varrho_n\mathbf{u}_n)(0+) \to \mathbf{m}_0 \quad \text{in } L^1(\Omega),$$

$$\frac{|\varrho_n\mathbf{u}_n(0+)|^2}{\varrho_n(0+)} \to \frac{|\mathbf{m}_0|^2}{\varrho_0} \quad \text{in } L^1(\Omega),$$

and

$$\varrho_n(0+)Q(\vartheta_n(0+)) \to \chi_0 \quad \text{in } L^1(\Omega).$$

Finally, let κ and Q obey hypotheses (3.3), (3.4) and (3.5), (3.6) respectively.

Then the limit functions ϱ, $\mathbf{u}$ from the conclusion of Theorem 6.1 represent a variational solution of the thermal energy equation (1.36) in the sense of Definition 4.5 satisfying

$$\varrho Q(\vartheta)(0) = \chi_0,$$

with a non-negative temperature ϑ, which coincides with the weak limit $\overline{\vartheta}$ on the set where ϱ is positive, that is,

$$\varrho\overline{\vartheta} = \varrho\vartheta \quad \text{on } (0,T)\times\Omega.$$

Moreover, the initial energy for the limit problem is

$$E(0) = \int_\Omega \frac{1}{2}\frac{|\mathbf{m}|^2}{\varrho_0} + \varrho_0 P_\mathrm{e}(\varrho_0) + \chi_0 \, \mathbf{dx}.$$

Remark Note that in accordance with our discussion in Chapter 4 the initial values of ϱ, $\varrho\mathbf{u}$, and $\varrho Q(\vartheta)$ are well defined.

Proof (i) To begin with, observe that one can pass to the limit in the renormalized inequality (4.50). Indeed, by virtue of (6.69), (6.79), the functions ϑ_n converge to $\overline{\vartheta}$ a.a. on the set where ϱ is positive and, moreover,

$$\left.\begin{array}{r}\varrho_n Q_h(\vartheta_n) \to \varrho Q_h(\overline{\vartheta}) \\ \varrho_n Q_h(\vartheta_n)\mathbf{u}_n \to \varrho Q_h(\overline{\vartheta})\mathbf{u} \\ h(\vartheta_n)\vartheta_n \, p_\vartheta(\varrho_n)\mathrm{div}_x\, \mathbf{u}_n \to h(\overline{\vartheta})\overline{\vartheta}\, p_\vartheta(\varrho)\mathrm{div}_x\, \mathbf{u}\end{array}\right\} \text{ weakly in } L^1((0,T) \times \Omega)$$

for any h satisfying (4.48). See (4.49) for the exact definition of Q_h and κ_h.

On the other hand, we showed in Lemma 4.8 that

$$\int_0^T \int_\Omega h(\overline{\vartheta})\, \mathbb{S} : \nabla_x \mathbf{u} \,\varphi \,\mathrm{d}x\,\mathrm{d}t \leq \liminf_{n\to\infty} \int_0^T \int_\Omega h(\vartheta_n)\, \mathbb{S}_n : \nabla_x\mathbf{u}_n \,\varphi \,\mathrm{d}x\,\mathrm{d}t$$

for any non-negative test function φ.

Furthermore, as the functions h tend to zero for large arguments, we can use Proposition 2.1 in order to conclude that

$$\kappa_h(\vartheta_n) \to \overline{\kappa_h(\vartheta)} \text{ weakly in } L^1((0,T) \times \Omega)$$

for any function h as in (4.48), where

$$\varrho\overline{\kappa_h(\vartheta)} = \varrho\kappa_h(\overline{\vartheta}) \quad \text{on } (0,T) \times \Omega.$$

Finally, in accordance with our hypotheses, we discover the limit

$$\int_\Omega \varrho_n(0+)Q_h(\vartheta_{0,n})\varphi(0)\,\mathrm{d}\mathbf{x} \to \int_\Omega \varrho_0 Q_h(\vartheta_0)\varphi(0)\,\mathrm{d}\mathbf{x} \quad \text{for } n \to \infty$$

in order to conclude that ϱ, $\mathbf{u}$, and $\overline{\vartheta}$ satisfy

$$\int_0^T \int_\Omega \varrho Q_h(\overline{\vartheta})\, \partial_t\varphi + \varrho Q_h(\overline{\vartheta})\mathbf{u} \cdot \nabla_x\varphi + \overline{\kappa_h(\vartheta)}\Delta\varphi \,\mathrm{d}\mathbf{x}\,\mathrm{d}t$$

$$\leq \int_0^T \int_\Omega h(\overline{\vartheta})\Big(\overline{\vartheta}p_\vartheta(\varrho)\, \mathrm{div}_x\, \mathbf{u} - \mathbb{S} : \nabla_x\mathbf{u}\Big)\varphi \,\mathrm{d}\mathbf{x}\,\mathrm{d}t - \int_\Omega \varrho_0 Q_h(\vartheta_0)\,\mathrm{d}\mathbf{x} \quad (6.81)$$

for any test function φ as in (4.42).

(ii) The idea now is to take

$$h(\vartheta) = \frac{1}{(1+\vartheta)^\omega}, \quad 0 < \omega < 1,$$

in the renormalized inequality (6.81), and to let $\omega \to 0$ to obtain (4.41) for any test function φ as in (4.42). This is possible as

$$\frac{1}{(1+\vartheta)^\omega} \nearrow 1 \quad \text{for } \omega \to 0,$$

and the monotone convergence theorem for the Lebesgue integral can be used. Note that

$$\overline{\kappa_h(\vartheta)} \nearrow \overline{\kappa(\vartheta)},$$

where

$$\varrho\overline{\kappa(\vartheta)} = \varrho\kappa(\overline{\vartheta}) \quad \text{on } (0,T) \times \Omega,$$

$$\int_0^T \int_\Omega \overline{\kappa(\vartheta)} \, \mathrm{d}\mathbf{x} \, \mathrm{d}t \leq \liminf_{n\to\infty} \int_0^T \int_\Omega \kappa(\vartheta_n) \, \mathrm{d}\mathbf{x} \, \mathrm{d}t. \qquad (6.82)$$

Finally, we set

$$\vartheta \equiv \kappa^{-1}(\overline{\kappa(\vartheta)}).$$

Obviously, the new function ϑ is non-negative, satisfies (4.41) with

$$\varrho Q(\varrho)(0) = \chi_0$$

for any φ, and

$$\varrho\overline{\vartheta} = \varrho\vartheta \quad \text{on } (0,T) \times \Omega.$$

(iii) In order to complete the proof, we have to show that the limit functions satisfy the energy inequality (4.45). In accordance with our hypotheses, we have

$$E_n(\tau) = \int_\Omega \frac{1}{2} \frac{|\varrho_n \mathbf{u}_n(\tau)|^2}{\varrho_n(\tau)} + \varrho_n(\tau) P_e(\varrho_n(\tau)) + \varrho_n(\tau) Q(\vartheta_n(\tau)) \, \mathrm{d}\mathbf{x}$$

$$\leq \int_\Omega \frac{1}{2} \frac{|\varrho_n \mathbf{u}_n(0+)|^2}{\varrho_n(0+)} + \varrho_n(0+) P_e(\varrho_n(0+)) + \varrho_n(0+) Q(\vartheta_{0,n}) \, \mathrm{d}\mathbf{x}$$

$$+ \int_0^\tau \int_\Omega \varrho_n \mathbf{f}_n \cdot \mathbf{u}_n \, \mathrm{d}\mathbf{x} \, \mathrm{d}t,$$

where the right-hand side tends to

$$\int_\Omega \frac{1}{2} \frac{|\mathbf{m}_0|^2}{\varrho_0} + \varrho_0 P_e(\varrho_0) + \chi_0 \, \mathrm{d}\mathbf{x} + \int_0^\tau \int_\Omega \varrho\mathbf{f} \cdot \mathbf{u} \, \mathrm{d}\mathbf{x} \, \mathrm{d}t$$

for $n \to \infty$.

On the other hand, as the mechanical energy is weakly lower semi-continuous (see Section 4.2), we have

$$\int_\Omega \frac{1}{2} \frac{|\varrho \mathbf{u}(\tau)|^2}{\varrho(\tau)} + \varrho(\tau) P_e(\varrho(\tau))\, d\mathbf{x} \leq \liminf_{n\to\infty} \int_\Omega \frac{1}{2} \frac{|\varrho_n \mathbf{u}_n(\tau)|^2}{\varrho_n(\tau)} + \varrho_n(\tau) P_e(\varrho_n(\tau))\, d\mathbf{x}$$

for any $\tau \in (0, T)$.

Finally, since

$$\varrho_n Q(\vartheta_n) \to \varrho Q(\vartheta) \quad \text{in } L^1((0, T) \times \Omega),$$

we get

$$\int_\Omega \varrho_n(\tau) Q(\vartheta_n(\tau))\, d\mathbf{x} \to \int_\Omega \varrho(\tau) Q(\vartheta(\tau))\, d\mathbf{x} \quad \text{for a.a. } \tau \in (0, T),$$

which completes the proof of the energy inequality (4.45).

□

6.8 Limits of bounded sequences in L^1

6.8.1 Renormalized limits

The idea of passing to the limit in the terms $\kappa(\vartheta_n)$ deserves some comments. First of all, we only know that

$$\kappa(\vartheta_n) \text{ are bounded in } L^1((0, T) \times \Omega),$$

which is not enough for the sequence $\{\kappa(\vartheta_n)\}_{n=1}^\infty$ to be precompact in the weak topology of the Lebesgue space $L^1((0, T) \times \Omega)$.

Thus we approximated the function κ by a family of functions κ_h such that

$$\kappa_h \nearrow \kappa \text{ for } h \nearrow 1, \quad \lim_{z\to\infty} \frac{\kappa_h(z)}{\kappa(z)} = 0.$$

Now the sequence $\{\kappa_h(\vartheta_n)\}_{n=1}^\infty$ is equi-integrable, and one can use Proposition 2.1 to obtain

$$\kappa_h(\vartheta_n) \to \overline{\kappa_h(\vartheta)} \text{ weakly in } L^1((0, T) \times \Omega) \quad \text{for } n \to \infty$$

for any fixed h.

Finally, as weak convergence is order preserving, we can use the monotone convergence theorem (Levi's theorem) to *define*

$$\overline{\kappa(\vartheta)} \equiv \lim_{h\to 1} \overline{\kappa_h(\vartheta)} \in L^1((0, T) \times \Omega).$$

Transforming this procedure into a definition, we can introduce a concept of *renormalized limit* for sequences $\{v_n\}_{n=1}^\infty$ of non-negative functions which are merely bounded in $L^1(O)$, where $O \subset R^M$ is a bounded domain.

Consider a sequence of continuous functions

$$\chi_k : [0, \infty) \to [0, \infty), \quad \chi_k(z) \nearrow z \quad \text{if } k \to \infty \quad \text{for any } z \geq 0,$$

such that

$$\lim_{z \to \infty} \frac{\chi_k(z)}{z} = 0 \quad \text{for any } k = 1, 2, \ldots . \tag{6.83}$$

In accordance with Proposition 2.1, we can secure a subsequence such that

$$\chi_k(v_n) \to \overline{\chi_k(v)} \quad \text{weakly in } L^1(O)$$

for any $k = 1, 2, \ldots$.

Now, we define a *renormalized limit* $v \in L^1(O)$ of $\{v_n\}_{n=1}^\infty$,

$$v = \mathrm{r} - \lim_{n \to \infty} v_n$$

as the unique limit

$$v(\mathbf{y}) = \lim_{k \to \infty} \overline{\chi_k(v)}(\mathbf{y}) \quad \text{for a.a. } \mathbf{y} \in O.$$

Lemma 6.4 *The renormalized limit is independent of the choice of the approximating sequence $\{\chi_k\}_{k=1}^\infty$.*

Proof   Consider two sequences $\{\chi_k^1\}_{k=1}^\infty$, $\{\chi_k^2\}_{k=1}^\infty$ yielding the renormalized limits v^1, v^2 respectively. Obviously, it is enough to show that

$$v^1 \geq \overline{\chi_m^2(v)} \quad \text{for any } m = 1, 2, \ldots .$$

Let $\varepsilon > 0$ be given. As χ_m^2 satisfies hypothesis (6.83), there exists $M = M(\varepsilon)$ such that

$$\int_{\{v_n \geq M\}} \chi_m^2 \, \mathrm{d}\mathbf{y} < \varepsilon \quad \text{for all } n = 1, 2, \ldots .$$

On the other hand, as the convergence $\chi_k^1(z) \nearrow z$ is uniform on $[0, M]$, there exists $k_0 = k_0(\varepsilon)$ such that

$$\chi_k^1(z) \geq \chi_m^2(z) - \frac{\varepsilon}{|O|} \quad \text{for any } z \in [0, M] \text{ and } k \geq k_0.$$

Thus for any measurable set $B \subset O$, we get

$$\int_B \chi_k^1(v_n) \, \mathrm{d}\mathbf{y} = \int_{B \cap \{v_n \leq M\}} \chi_k^1 \, \mathrm{d}\mathbf{y} + \int_{B \cap \{v_n > M\}} \chi_k^1 \, \mathrm{d}\mathbf{y}$$

$$\geq \int_{B \cap \{v_n \leq M\}} \chi_m^2(v_n) \, \mathrm{d}\mathbf{y} - \varepsilon$$

$$\geq \int_B \chi_m^2(v_n) \, \mathrm{d}\mathbf{y} - \int_{B \cap \{v_n > M\}} \chi_m^2(v_n) \, \mathrm{d}\mathbf{y} - \varepsilon$$

$$\geq \int_B \chi_m^2(v_n) \, \mathrm{d}\mathbf{y} - 2\varepsilon$$

provided $k \geq k_0$.

This yields

$$v^1 \geq \overline{\chi_k^1(v)} \geq \overline{\chi_m^2(v)} - \varepsilon \quad \text{for } k \geq k_0.$$

As $\varepsilon > 0$ was arbitary, the proof is complete.

□

6.8.2 Biting limits

The above procedure yields a renormalized limit for a sequence of functions which is merely bounded in the (non-reflexive) space $L^1(O)$. There is another concept of a so-called *biting limit* based on the following result due to Chacon (see [16]).

Lemma 6.5 *Assume that $O \subset R^M$ is a bounded measurable set, and $\{v_n\}_{n=1}^{\infty}$ a bounded sequence in $L^1(O)$.*

Then there exists a function $v \in L^1(O)$, a subsequence $\{v_n\}_{n=1}^{\infty}$, and a non-increasing sequence of measurable sets $E_m \subset O$,

$$|E_m| \to 0 \quad \text{for } m \to \infty,$$

such that

$$v_n \to v \text{ weakly in } L^1(O \setminus E_m) \quad \text{as } n \to \infty$$

for any fixed m.

The function v is termed a biting limit of the sequence $\{v_n\}_{n=1}^{\infty}$.

An elementary proof of Lemma 6.5 was given by Ball and Murat (see Lemma in [7]), where the authors also showed the following result (see Proposition in [7]).

Proposition 6.4 *Let $\chi_k \in C(R)$ be a family of continuous functions such that*

$$\chi_k(z) \to z \text{ as } k \to \infty \quad \text{for any } z \in R,$$

and

$$\lim_{|z| \to \infty} \frac{|\chi_k(z)|}{|z|} = 0 \quad \text{for each } k.$$

Furthermore, let $\{v_n\}_{n=1}^{\infty}$ be a bounded sequence in $L^1(O)$ possessing a biting limit v.

Then for each k we have

$$\chi_k(v_n) \to \overline{\chi_k(v)} \text{ weakly in } L^1(O) \quad \text{as } n \to \infty,$$

and

$$\overline{\chi_k(v)} \to v \text{ (strongly) in } L^1(O) \quad \text{as } k \to \infty.$$

In particular, the renormalized limit introduced above and the biting limit coincide (modulo a subsequence).

In light of Proposition 6.4, the reader will have noticed that the assumption of monotone convergence of the functions $\{\chi_k\}_{k=1}^{\infty}$ in the definition of the renormalized limit is not necessary. By the same token, the functions v_n need not be non-negative.

6.9 Bibliographical notes

6.1–6.2 The weak continuity of the effective viscous pressure as stated in Proposition 6.1 is a remarkable result of P.-L. Lions [85]. In this context, it is interesting to note that some regularity properties of this quantity were also observed by Hoff [64]. Assuming *a priori* validity of Proposition 6.1, Serre [108] described possible oscillations of the density in terms of the corresponding Young measure in a spirit similar to Section 6.6.

The weak continuity of the effective viscous pressure reveals an intimate relation between the continuity equation (1.34) and the momentum equation (1.35). We should always keep in mind that this property leans essentially on the hypothesis that the viscosity coefficients μ and λ are constant. On the other hand, if this is not the case, weak sequential stability of the problem can be saved with the help of more classical methods. Vaigant and Kazhikhov [119] considered the isentropic flow in two space dimensions with μ constant, and $\lambda \approx \varrho^\beta$, $\beta \geq 3$. Even though these hypotheses are probably physically irrelevant, the mathematical theory they develop yields the existence and uniqueness of global in-time *regular* solutions of the problem.

6.3–6.5 The oscillations defect measure was introduced in [45] in order to extend the existence theory for isentropic flows developed in [85] to the full range of the adiabatic constants $\gamma > 1$ for $N = 2$ and $\gamma > \frac{3}{2}$ for $N = 3$. The idea of investigating boundedness of a weak limit of a difference of two nonlinear quantities rather than treating them separately was also used by Jiang and Zhang in [67].

6.6 The fact that density oscillations cannot be created in a finite time provided the pressure p is a non-decreasing function of the density represents the heart of the existence theory presented in [85]. A general (non-monotone) barotropic pressure law was treated in [47]. Uniform-in-time decay of oscillations for a convex pressure was shown in [53].

The fact that oscillations decay at a uniform rate plays an essential role when studying the asymptotic properties of solutions for large values of time. In terms of the theory of dynamical systems, one can say that the ω-limit set of the density trajectory $\{\varrho(t)\}_{t \geq 0}$ is precompact with respect to the *strong* topology of the space $L^1(\Omega)$. This property, together with the existence of bounded absorbing sets was used to develop a dynamical systems theory for the Navier–Stokes system describing a barotropic flow (see e.g. [44, 46]).

6.7 One of the first versions of the Aubin–Lions lemma can be found in [81]. Several generalizations were obtained by Simon [109]. Here we adopted a version proved in [54].

6.8 Several attempts have been made to give a precise meaning to concentration sets. A nice survey of different concepts of "defect measures" is included in [41].

DiPerna and Majda [33] introduced a reduced defect measure

$$m(B) = \limsup_{n \to \infty} \int_B |v_n - v|^2 \, \mathrm{d}\mathbf{x}$$

for any Borel set B in order to study concentration sets for solutions of the Euler equations.

The definition of the renormalized limit introduced in this book is motivated by the aforementioned paper by Ball and Murat [7]. Their approach is strongly related to the concept of Young measures developed in the context of the compensated compactness theory (see e.g. [6, 111]).

7

GLOBAL EXISTENCE

The principal criterion of applicability of any *mathematical model* is its *well-posedness* According to Hadamard, this issue comprises a thorough discussion of the following topics:

(1) *Existence of solutions for given data.* The *data* for our problem are usually the values of the *macroscopic state variables* ϱ, $\varrho\mathbf{u}$, and $\varrho Q(\vartheta)$ at the initial time $t = 0$; the driving force $\mathbf{f}$, and other quantities as the case may be. The problem is whether or not there exists a solution of the Navier–Stokes system for any choice of data.

(2) *Uniqueness.* Any mathematical model is to be *deterministic* specifically, the state of the system at any future time $t > 0$ should be uniquely determined by its state at $t = 0$.

(3) *Stability.* Small perturbations of the data should result in small variation of the corresponding solution at least on a given (compact) time interval.

To begin with, let us say honestly that a complete answer to these questions for the *Navier–Stokes system* (1.34)–(1.36) is far from being settled. The partial results presented in this book should only shed some light on the amazing complexity of the problem.

Having collected all the necessary material we are in a position to address the main topic of this book—a rigorous *existence theory* for the Navier–Stokes system describing the motion of a viscous, compressible, and heat-conducting fluid. The problem is to establish the existence of three macroscopic quantities: the density ϱ, the velocity $\mathbf{u}$, and the temperature ϑ solving the equation of continuity (1.34), the momentum equation (1.35), and the thermal energy equation (1.36) with the boundary conditions (1.41), (1.42), and the initial conditions (1.43) on a given spatial domain Ω, and a given time interval $(0, T)$.

Recall from the earlier exposition that we do not expect to find classical (smooth) solutions for any given data but, on the other hand, we present a rather complete theory within the framework of the *variational solutions* introduced in Chapter 4. For variational solutions, the instantaneous values of the macroscopic variables are characterized by their distribution (measures) over the spatial domain Ω, and the differential equations (1.34)–(1.36) are replaced by a family of integral identities related to the three fundamental principles of continuum mechanics: conservation of mass, momentum, and total energy.

Although the variational solutions could be very far from the standard notion of a solution to a differential equation, they still meet two basic principles of *compatibility* with the system (1.34)–(1.36): (i) any smooth (classical) solution of the problem is a variational solution; (ii) any variational solution which is smooth is a classical solution.

7.1 Statement of the main result

In order to avoid unnecessary technical problems, we shall assume that $\Omega \subset R^N$, $N \geq 2$ is a bounded domain with (regular) boundary of class $C^{2+\nu}$, $\nu > 0$; and we fix the time interval to be $(0, T)$.

Definition 7.1 We shall say that a triple of functions ϱ, $\mathbf{u}$, and ϑ is a *variational solution* of the *Navier–Stokes system* (1.34)–(1.36) supplemented with the boundary conditions (1.41), (1.42), and the initial conditions (1.43) if

(1) the density $\varrho = \varrho(t, \mathbf{x})$ and the velocity $\mathbf{u} = \mathbf{u}(t, \mathbf{x})$ represent a variational solution of the *equation of continuity* (1.34) in the sense of Definition 4.2;

(2) the functions ϱ, $\mathbf{u}$, and ϑ solve the *momentum equation* (1.35) in the sense of Definition 4.3, where the pressure p and the viscous stress tensor $\mathbb{S}$ are given by the constitutive equations (1.39) and (1.37), respectively;

(3) the temperature $\vartheta = \vartheta(t, \mathbf{x})$ together with ϱ and $\mathbf{u}$ form a variational solution of the *thermal energy equation* (1.36) in the sense of Definition 4.5;

(4) the functions ϱ, $\varrho\mathbf{u}$, and $\varrho Q(\vartheta)$ satisfy the *initial conditions*

$$\left\{ \begin{array}{l} \text{ess}\lim_{t \to 0+} \int_\Omega \varrho\eta \, d\mathbf{x} = \int_\Omega \varrho_0\eta \, d\mathbf{x}, \\[2mm] \text{ess}\lim_{t \to 0+} \int_\Omega (\varrho\mathbf{u})(t) \cdot \eta \, d\mathbf{x} = \int_\Omega \mathbf{m}_0 \cdot \eta \, d\mathbf{x}, \\[2mm] \text{ess}\lim_{t \to 0+} \int_\Omega \varrho Q(\vartheta)(t)\eta \, d\mathbf{x} = \int_\Omega \chi_0\eta \, d\mathbf{x} \end{array} \right\}$$

for any $\eta \in \mathcal{D}(\Omega)$.

In accordance with our discussion in Chapter 4, any variational solution ϱ, $\mathbf{u}$, and ϑ enjoys the following properties:

(1) the equation of continuity

$$\partial_t\varrho + \text{div}_x(\varrho\mathbf{u}) = 0,$$

the momentum equation

$$\partial_t(\varrho\mathbf{u}) + \mathrm{div}_x(\varrho\mathbf{u}\otimes\mathbf{u}) + \nabla_x\Big(p_e(\varrho) + \vartheta p_\vartheta(\varrho)\Big) = \mathrm{div}_x\mathbb{S} + \varrho\mathbf{f},$$

and the thermal energy inequality

$$\partial_t\big(\varrho Q(\vartheta)\big) + \mathrm{div}_x(\varrho Q(\vartheta)\mathbf{u}) - \Delta\mathcal{K}(\vartheta) \geq \mathbb{S}:\nabla_x\mathbf{u} - \vartheta p_\vartheta(\varrho)\,\mathrm{div}_x\mathbf{u}$$

hold in $\mathcal{D}'((0,T)\times\Omega)$;

(2) the total *energy inequality*

$$E(\tau) \equiv \int_\Omega \frac{1}{2}\frac{|\varrho\mathbf{u}(\tau)|^2}{\varrho(\tau)} + \varrho(\tau)P_e(\varrho(\tau)) + \varrho Q(\vartheta)(\tau)\,\mathrm{d}\mathbf{x}$$

$$\leqslant E(0) + \int_0^\tau \int_\Omega \varrho\mathbf{f}\cdot\mathbf{u}\,\mathrm{d}\mathbf{x}\,\mathrm{d}t$$

holds for a.a. $\tau\in[0,T]$, with

$$E(0) = \int_\Omega \frac{1}{2}\frac{|\mathbf{m}|^2}{\varrho_0} + \varrho_0 P_e(\varrho_0) + \chi_0\,\mathrm{d}\mathbf{x};$$

(3) the instantaneous value of the density (see Definition 2.1)

$$t\mapsto \begin{cases} \varrho(t+) & \text{for } t=0, \\ \varrho(t) & \text{if } t\in[0,T], \\ \varrho(t-) & \text{for } t=T \end{cases}$$

is a continuous function of $t\in[0,T]$ with respect to the topology of the space

$$L^1(\Omega)\cap L^\gamma_{\text{weak}}(\Omega),$$

the *total mass*

$$M = \int_\Omega \varrho(t)\,\mathrm{d}\mathbf{x} \text{ is a constant of motion;}$$

(4) the velocity $\mathbf{u}(t)$ belongs to the Sobolev space $W_0^{1,2}(\Omega)$ for a.a. $t\in(0,T)$, in particular, $\mathbf{u}$ satisfies the boundary conditions (1.41) in the sense of traces;

(5) the instantaneous value of the momentum

$$t \mapsto \begin{cases} \varrho\mathbf{u}(t+) & \text{for } t = 0, \\ \varrho\mathbf{u}(t) & \text{if } t \in (0, T), \\ \varrho\mathbf{u}(t-) & \text{for } t = T \end{cases}$$

is a continuous function $t \in [0, T]$ with respect to the topology of the space

$$L_{\text{weak}}^{m_\infty}(\Omega; R^N) \quad \text{with } m_\infty > 1;$$

(6) the temperature ϑ satisfies the homogeneous Neumann boundary conditions (1.42) in the sense of Definition 4.5 (that is, through the choice of test functions in (4.41)), moreover, ϑ is positive a.a. on $(0, T) \times \Omega$, more specifically,

$$\log(\varrho) \in L^2((0, T) \times \Omega);$$

(7) the instantaneous value of $\varrho Q(\vartheta)(\tau)$ is a non-negative measure which is absolutely continuous with respect to the standard Lebesgue measure on Ω for a.a. $\tau \in (0, T)$, and

$$\varrho Q(\vartheta)(\tau+) \geq \varrho Q(\vartheta)(\tau-) \quad \text{for any } \tau \in (0, T),$$
$$\varrho Q(\vartheta)(\tau) \rightarrow \chi_0 \text{ in } \mathcal{M}(\overline{\Omega}) \quad \text{for } \tau \rightarrow 0+ .$$

For *barotropic flows* where $p = p_e(\varrho)$, the total energy inequality reduces to the standard *mechanical energy inequality*:

$$E_{\text{mech}}(\tau) + \int_0^\tau \int_\Omega \mathbb{S} : \nabla_x \mathbf{u} \, \mathrm{d}\mathbf{x} \, \mathrm{d}t \leq E_{\text{mech}}(0) + \int_0^\tau \int_\Omega \varrho\mathbf{u} \cdot \mathbf{f} \, \mathrm{d}\mathbf{x} \, \mathrm{d}t,$$

where

$$E_{\text{mech}}(\tau) \equiv \int_\Omega \frac{1}{2} \frac{|\varrho\mathbf{u}(\tau)|^2}{\varrho(\tau)} + \varrho(\tau) P_e(\varrho(\tau)) \, \mathrm{d}\mathbf{x},$$

$$E_{\text{mech}}(0) \equiv \int_\Omega \frac{1}{2} \frac{|\mathbf{m}|^2}{\varrho_0} + \varrho_0 P_e(\varrho_0) \, \mathrm{d}\mathbf{x}.$$

Consequently, from the purely mathematical viewpoint, the *barotropic system* may be regarded as a particular case of the full system (1.34)–(1.36) though their respective physical backgrounds are rather different.

Our main result reads as follows.

Theorem 7.1 *Let $\Omega \subset R^N$, $N \geq 2$, be a bounded domain with boundary of class $C^{2+\nu}$, $\nu > 0$. Suppose that the pressure p is given by the constitutive equation (1.39), where the functions $p_e = p_e(\varrho)$ and $p_\vartheta = p_\vartheta(\varrho)$ comply with hypotheses (3.1), (3.2) with*

$$\gamma > \frac{N}{2}. \tag{7.1}$$

Furthermore, assume that the viscous stress tensor $\mathbb{S}$ is given by (1.37) with constant viscosity coefficients μ and λ satisfying (1.38). Suppose the heat flux $\mathbf{q}$ is determined by Fourier's law (1.40), where κ satisfies hypotheses (3.3), (3.4). Finally, let Q satisfy (3.5) and (3.6).
As for the data, suppose that

$$\left\{ \begin{array}{c} \varrho_0 \geq 0, \quad \varrho_0 \in L^\gamma(\Omega), \\[2mm] \dfrac{|\mathbf{m}_0|^2}{\varrho_0} \in L^1(\Omega), \\[2mm] \chi_0 = \varrho_0 Q(\vartheta_0), \quad \text{with } \vartheta_0 \in L^\infty(\Omega), \ \text{ess}\inf_\Omega \vartheta_0 > 0. \end{array} \right\} \tag{7.2}$$

Moreover, let the driving force $\mathbf{f}$ be a bounded measurable function on $(0, T) \times \Omega$.
   Then the problem (1.34)–(1.36) supplemented by the boundary conditions (1.41), (1.42), and the initial conditions (1.43) possesses at least one variational solution ϱ, $\mathbf{u}$, ϑ on $(0, T) \times \Omega$ in the sense of Definition 7.1.

Remark   Note that the hypothesis of integrability of the initial kinetic energy

$$\frac{1}{2}\frac{|\mathbf{m}_0|^2}{\varrho_0} \in L^1(\Omega)$$

requires implicitly the compatibility condition

$$\mathbf{m}_0(\mathbf{x}) = 0 \quad \text{for a.a. } \mathbf{x} \in \{\varrho_0 = 0\}$$

to hold.

The rest of this chapter is devoted to the proof of Theorem 7.1, the main ideas being, of course, those presented in Chapter 6. In particular, the "weak continuity" of the *effective viscous pressure* stated in Proposition 6.1, boundedness of the *oscillations defect measure* $\text{osc}[\varrho_n \to \varrho]$ proved in Proposition 6.2, and the *propagation of oscillations* results discussed in Section 6.6 will be used. Moreover, in order to handle the thermal energy equation (1.36), we adapt the *a priori* estimates of the temperature obtained in Section 5.2 together with the concept of renormalized solutions of (1.36) introduced in Definition 4.4. The problem of low integrability of the *thermal conductivity potential* $\mathcal{K}(\vartheta)$ will be solved by means of the technique based on *renormalized limits* discussed in Section 6.7.1.

In the light of available *a priori* estimates of the velocity **u** (cf. Section 3), the condition (7.1) on γ seems almost optimal as it renders the convective term $\varrho\mathbf{u} \otimes \mathbf{u}$ integrable. On the other hand, the corresponding hypothesis (3.2) for the thermal pressure p_ϑ seems much more restrictive and could probably be removed at the expense of a more severe stipulation concerning the admissible growth of the heat conduction coefficient κ postulated in (3.4). Note, however, that the "adiabatic coefficient" $\gamma = 3$, which allows for linear growth of p_ϑ, appears, for instance, in the theory of nuclear fluids studied in [34, 35]. The condition (3.4) itself is physically reasonable as experiments predict the value of $\alpha \approx 4.5$–5.5 while Q should behave like $\vartheta^{1.5}$ for large arguments, which is in a good agreement with hypothesis (3.6) (see e.g. [126, 10]).

Probably the most disputable feature of our model from the physical viewpoint is the assumption that the viscosity coefficients μ and λ are constant in the whole range of ϱ and ϑ. Relaxation of this rather unnatural hypothesis would require, however, a completely different mathematical approach not available at the present time.

7.2 The approximation scheme

There are several possibilities of how to construct the variational solutions, the existence of which is claimed in Theorem 7.1. Here, we employ a three level *approximation scheme* based on solving the following system of equations:

---

continuity equation with vanishing viscosity:

$$\partial_t\varrho + \operatorname{div}_x(\varrho\mathbf{u}) = \varepsilon\Delta\varrho \quad \text{on } (0,T) \times \Omega, \ \varepsilon > 0, \tag{7.3}$$

with the homogeneous Neumann boundary condition

$$\nabla_x\varrho \cdot \mathbf{n} = 0 \quad \text{on } \partial\Omega, \tag{7.4}$$

and the initial condition

$$\varrho(0) = \varrho_{0,\delta} \quad \text{on } \Omega; \tag{7.5}$$

---

momentum equation with artificial pressure:

$$\partial_t(\varrho\mathbf{u}) + \mathrm{div}_x(\varrho\mathbf{u}\otimes\mathbf{u}) + \nabla_x\Big(p(\varrho,\vartheta)+\delta\varrho^\beta\Big) + \varepsilon\nabla_x\mathbf{u}\nabla_x\varrho$$

$$= \mathrm{div}_x\mathbb{S} + \varrho\mathbf{f} \quad \text{on } (0,T)\times\Omega, \quad \delta>0, \quad \beta>1, \qquad (7.6)$$

with

$$\mathbf{u}=0 \quad \text{on } \partial\Omega, \qquad (7.7)$$

$$(\varrho\mathbf{u})(0) = \mathbf{m}_{0,\delta} \quad \text{on } \Omega; \qquad (7.8)$$

---

and

---

regularized thermal energy equation:

$$\partial_t\Big((\delta+\varrho)Q(\vartheta)\Big) + \mathrm{div}_x(\varrho Q(\vartheta)\mathbf{u}) - \Delta\mathcal{K}(\vartheta) + \delta\vartheta^{\alpha+1}$$

$$= (1-\delta)\,\mathbb{S}:\nabla_x\mathbf{u} - \vartheta p_\vartheta\,\mathrm{div}_x\mathbf{u} \quad \text{on } (0,T)\times\Omega, \qquad (7.9)$$

with

$$\nabla_x\vartheta\cdot\mathbf{n}=0 \quad \text{on } \partial\Omega, \qquad (7.10)$$

$$(\delta+\varrho)Q(\vartheta)(0,\mathbf{x}) = \chi_{0,\delta}(\mathbf{x}) \equiv (\delta+\varrho_{0,\delta})Q(\vartheta_{0,\delta}) \quad \text{on } \Omega. \qquad (7.11)$$

The extra term $\varepsilon\Delta\varrho$ in (7.3) represents a *"vanishing viscosity"* with no specific physical meaning. From the mathematical viewpoint, however, it converts the hyperbolic equation (1.34) into a parabolic one. As a result, one can expect better regularity properties of the "densities" ϱ constructed at this level of approximation.

The quantity $\delta\varrho^\beta$ added to the momentum equation (1.35) can be considered as an *artificial pressure*, which was introduced to make the pressure estimates presented in Chapter 5 compatible with the vanishing viscosity regularization of (1.34). More precisely, the pressure estimates based on multiplication of (1.35) by the quantity $\nabla_x\Delta^{-1}\varrho^\omega$ will still remain in force for the modified system (7.3), (7.6) only if $\omega=1$. Accordingly, one must take $\beta=\beta(N)$ large enough to be able to exploit the ideas of the proof of Proposition 5.1.

In the same spirit, the new quantity $\varepsilon\nabla_x\mathbf{u}\nabla_x\varrho$ was introduced in (7.6) in order to eliminate the extra terms arising in the energy inequality to save the *a priori estimates* obtained in Chapter 3.

Finally, there are several terms in the regularized thermal energy equation (7.9) depending on the (positive) parameter δ. The reason for introducing these quantities is simple—to avoid technicalities connected with the temperature estimates presented in Chapter 5 at the first level of approximation.

The initial data are modified as follows:

(1) The density $\varrho_{0,\delta} \in C^{2+\nu}(\overline{\Omega})$, $\nu > 0$, satisfies the homogeneous Neumann boundary condition

$$\nabla \varrho_{0,\delta} \cdot \mathbf{n}|_{\partial\Omega} = 0. \tag{7.12}$$

Furthermore, we suppose

$$0 < \delta \leq \varrho_{0,\delta}(\mathbf{x}) \leq \delta^{-1/2\beta} \quad \text{for all } \mathbf{x} \in \Omega, \tag{7.13}$$

$$\varrho_{0,\delta} \to \varrho_0 \text{ in } L^\gamma(\Omega), \quad \left|\{\mathbf{x} \in \Omega|\ \varrho_{0,\delta}(\mathbf{x}) < \varrho_0(\mathbf{x})\}\right| \to 0 \text{ as } \delta \to 0. \tag{7.14}$$

(2) The initial momenta $\mathbf{m}_{0,\delta}$ are defined as

$$\mathbf{m}_{0,\delta}(\mathbf{x}) = \begin{cases} \mathbf{m}_0 & \text{if } \varrho_{0,\delta}(\mathbf{x}) \geq \varrho_0(\mathbf{x}), \\ 0 & \text{for } \varrho_{0,\delta}(\mathbf{x}) < \varrho_0(\mathbf{x}). \end{cases} \tag{7.15}$$

(3) The functions $\vartheta_{0,\delta} \in C^{2+\nu}(\overline{\Omega})$ satisfy

$$0 < \underline{\vartheta} \leq \vartheta_{0,\delta}(\mathbf{x}) \leq \overline{\vartheta} \quad \text{for all } \mathbf{x} \in \Omega, \ \delta > 0, \ \nabla\vartheta_{0,\delta} \cdot \mathbf{n}|_{\partial\Omega} = 0. \tag{7.16}$$

Furthermore,

$$\vartheta_{0,\delta} \to \vartheta_0 \text{ in } L^1(\Omega) \quad \text{as } \delta \to 0.$$

In particular, the initial value of the (modified) total energy

$$E(0) = E_\delta(0) \equiv \int_\Omega \frac{1}{2} \frac{|\mathbf{m}_{0,\delta}|^2}{\varrho_{0,\delta}} + \varrho_{0,\delta} P_e(\varrho_{0,\delta}) + \varrho_{0,\delta} Q(\vartheta_{0,\delta}) + \frac{\delta}{\beta - 1} \varrho_{0,\delta}^\beta \, \mathrm{d}\mathbf{x} \tag{7.17}$$

is bounded by a constant independent of $\delta > 0$.

The principal strategy of the proof of Theorem 7.1 will be to solve first the full system (7.3)–(7.11) for positive values of the parameters ε and δ; then to let $\varepsilon \to 0$ to get rid of the artificial viscosity in (7.3); and finally, evoking the full strength of the pressure and temperature estimates obtained in Chapter 5, we pass to the limit for $\delta \to 0$ to recover the original system (1.34)–(1.36).

7.3 The Faedo–Galerkin approximations

In spite of the fact that the system (7.3)–(7.11) is of parabolic type, a suitable existence result is hard to find in the literature. This is mainly because the unknowns $\mathbf{u}$ and ϑ appear to be multiplied by ϱ in the leading terms of (7.6), (7.9) respectively, which puts the problem out of the standard setting. Consequently, we are forced to use a more complicated approach based on the *Faedo–Galerkin approximation* technique to obtain the first level approximate solutions.

Consider a finite-dimensional (Hilbert) space

$$X_n = \text{span}\{\eta_j\}_{j=1}^n,$$

where $\eta_j \in \mathcal{D}(\Omega)^N$ are linearly independent vector functions ranging in the N-dimensional Euclidean space R^N.

The approximate velocities $\mathbf{u}_n \in C([0,T]; X_n)$ are looked for to satisfy an integral identity

$$\int_\Omega \varrho \mathbf{u}_n(t) \cdot \eta \, d\mathbf{x} - \int_\Omega \mathbf{m}_{0,\delta} \cdot \eta \, d\mathbf{x}$$

$$= \int_0^t \int_\Omega \left[[\varrho \mathbf{u}_n \otimes \mathbf{u}_n] - \mathbb{S}_n \right] : \nabla_x \eta + \left[p(\varrho, \vartheta) + \delta \varrho^\beta \right] \text{div}_x \eta \, d\mathbf{x} \, ds$$

$$+ \int_0^t \int_\Omega [\varrho \mathbf{f} - \varepsilon \nabla_x \mathbf{u}_n \nabla_x \varrho] \cdot \eta \, d\mathbf{x} \, ds \qquad (7.18)$$

with

$$\mathbb{S}_n \equiv \mu(\nabla_x \mathbf{u}_n + \nabla_x \mathbf{u}_n^t) + \lambda \, \text{div}_x \mathbf{u}_n \, \mathbb{I}$$

for any test function $\eta \in X_n$, and all $t \in [0,T]$. The system (7.18) can be interpreted as a projection of the *infinite-dimensional dynamical system* represented by equation (7.6) onto a *finite system* of differential equations on X_n.

The *density*

$$\varrho = \varrho[\mathbf{u}_n]$$

appearing in the integral on the right-hand side of (7.18) is determined as the unique solution of the problem (7.3)–(7.5) with $\mathbf{u}$ replaced by $\mathbf{u}_n$. Similarly, the *temperature*

$$\vartheta = \vartheta[\mathbf{u}_n]$$

is defined through (7.9)–(7.11) with fixed $\mathbf{u} = \mathbf{u}_n$, and $\varrho = \varrho[\mathbf{u}_n]$.

Solvability of these auxilliary problems will be discussed in the next two sections.

7.3.1 On the regularized continuity equation

We first recall without proofs the basic facts from the theory of parabolic equations related to the linear problem:

$$\left\{ \begin{array}{c} \partial_t \varrho + \text{div}_x(\varrho \mathbf{u}) = \varepsilon \Delta \varrho \quad \text{on } (0,T) \times \Omega, \\ \nabla_x \varrho \cdot \mathbf{n}|_{\partial\Omega} = 0, \\ \varrho(0) = \varrho_{0,\delta} \end{array} \right\} \qquad (7.19)$$

with $\mathbf{u} \in C([0,T]; C^2(\overline{\Omega}; R^N))$ a given function.

Any regular solution of (7.19) satisfies the classical *a priori* estimates:

$$\|\partial_t \varrho\|_{C([0,T];C^\nu(\overline{\Omega}))} + \|\varrho\|_{C([0,T];C^{2+\nu}(\overline{\Omega}))}$$

$$\leq c(\varepsilon,\nu)\Big(\|\varrho_{0,\delta}\|_{C^{2+\nu}(\overline{\Omega})} + \|\mathrm{div}_x(\varrho\mathbf{u})\|_{C([0,T];C^\nu(\overline{\Omega}))}\Big), \quad \nu > 0 \qquad (7.20)$$

provided $\Omega \subset R^N$ is a bounded domain of class $C^{2+\nu}$, $\nu > 0$.

Moreover, the $(L^p - L^q)$ version of these estimates holds true:

$$\|\partial_t \varrho\|_{L^p(0,T;L^q(\Omega))} + \|\varrho\|_{L^p(0,T;W^{2,q}(\Omega))}$$

$$\leq c(\varepsilon,p,q)\Big(\|\varrho_{0,\delta}\|_{W^{2,q}(\Omega)} + \|\mathrm{div}_x(\varrho\mathbf{u})\|_{L^p(0,T;L^q(\Omega))}\Big), \quad \text{for any } 1 < p, q < \infty. \qquad (7.21)$$

Both (7.20) and (7.21) are derived from the corresponding estimates for the *Laplace operator* Δ supplemented by the homogeneous Neumann boundary conditions (see e.g. [2] for much more general classes of *elliptic operators* and boundary conditions).

Futhermore, solutions of *scalar parabolic equations* of second order obey the *maximum principle* (see e.g. [106]). In particular, as (7.19) is linear with respect to ϱ, we have

$$\big(\inf_{\mathbf{x}\in\Omega} \varrho(0,\mathbf{x})\big) \exp\Big(-\int_0^t \|\mathrm{div}_x \mathbf{u}_n(s)\|_{L^\infty(\Omega)}\, \mathrm{d}s\Big) \leq \varrho(t,x)$$

$$\leq \big(\sup_{\mathbf{x}\in\Omega} \varrho(0,\mathbf{x})\big) \exp\Big(\int_0^t \|\mathrm{div}_x \mathbf{u}_n(s)\|_{L^\infty(\Omega)}\, \mathrm{d}s\Big) \text{ for all } t \in [0,T], \ \mathbf{x} \in \Omega. \qquad (7.22)$$

Indeed the spatially homogeneous functions

$$\underline{\varrho}(t) \equiv \inf_{\mathbf{x}\in\Omega} \varrho(t,\mathbf{x}), \quad \overline{\varrho}(t) \equiv \sup_{\mathbf{x}\in\Omega} \varrho(t,\mathbf{x})$$

solve the problem

$$\partial_t \underline{\varrho}(t) + \nabla_x \underline{\varrho}(t) \cdot \mathbf{u} = \varepsilon \Delta \underline{\varrho}(t) - \underline{\varrho}(t)\|\mathrm{div}_x \mathbf{u}(t)\|_{L^\infty(\Omega)},$$

$$\nabla_x \underline{\varrho} \cdot \mathbf{n}|_{\partial\Omega} = 0,$$

and

$$\partial_t \overline{\varrho}(t) + \nabla_x \overline{\varrho}(t) \cdot \mathbf{u} = \varepsilon \Delta \overline{\varrho}(t) + \overline{\varrho}(t)\|\mathrm{div}_x \mathbf{u}_n(t)\|_{L^\infty(\Omega)},$$

$$\nabla_x \overline{\varrho} \cdot \mathbf{n}|_{\partial\Omega} = 0;$$

whence $\underline{\varrho}(t)$, $\overline{\varrho}(t)$ are respectively a sub and supersolution of (7.19).

The above estimates hold for smooth solutions of the problem. Turning now to the *weak solutions* we report the classical uniqueness result which can be found in [75].

Lemma 7.1 [Theorem 5.1 Chapter III, Section 5 in [75]] *Let* $\varrho_{0,\delta} \in L^2(\Omega)$ *be given.*

Then for any given $\mathbf{u} \in C([0,T]; C_0^2(\overline{\Omega}))$, *there is at most one weak solution* ϱ *of (7.19) belonging to the space* $L^2(0,T;W^{1,2}(\Omega))$.

Remark In accordance with the definition in Chapter III of Section 5 in [75], a function

$$\varrho \in L^2(0,T;W^{1,2}(\Omega))$$

is termed a *weak solution* of problem (7.19) if the integral identity

$$\int_0^T \int_\Omega \varepsilon \nabla_x \varrho \cdot \nabla_x \varphi - \varrho \partial_t \varphi \, \mathrm{d}\mathbf{x} \, \mathrm{d}t = \int_0^T \int_\Omega \varrho \mathbf{u} \cdot \nabla_x \varphi \, \mathrm{d}\mathbf{x} \, \mathrm{d}t + \int_\Omega \varrho_{0,\delta} \varphi \, \mathrm{d}\mathbf{x} \quad (7.23)$$

holds for any test function $\varphi \in C^{1,2}([0,T] \times \overline{\Omega})$ such that $\varphi(T) \equiv 0$ (cf. Definition 4.5).

On the other hand, there is a fairly satisfactory result about existence of *smooth solutions* of the problem (7.19).

Lemma 7.2 [Theorem 5.1.2 in [88]]
Let $\Omega \subset R^N$ *be a bounded domain of class* $C^{2+\nu}$, $\nu > 0$. *Assume that* $\varrho_{0,\delta} \in C^{2+\nu}(\overline{\Omega})$, $\mathbf{u} \in C([0,T]; C_0^2(\overline{\Omega}))$. *Moreover, let* $\nabla_x \varrho_{0,\delta} \cdot \mathbf{n} = 0$ *on* $\partial\Omega$.

Then the problem (7.19) possesses a unique classical solution ϱ *such that*

$$\partial_t \varrho \in C([0,T]; C^\nu(\overline{\Omega})), \quad \varrho \in C([0,T]; C^{2+\nu}(\overline{\Omega})).$$

Remark Unlike most of the existence theorems that can be found in classical monographs on parabolic equations, the above result does not require any regularity of the coefficient represented by the velocity field $\mathbf{u}$ with respect to time (with the exception of being continuous). This is very convenient here as the velocity $\mathbf{u}_n$ will be found with the help of a fixed point argument on the space $C([0,T]; X_n)$.

The information provided by estimates (7.20)–(7.22), and Lemmas 7.1, 7.2 can be recast in terms of the solution operator $\mathbf{u} \mapsto \varrho[\mathbf{u}]$.

Proposition 7.1 *Let* $\Omega \subset R^N$ *be a bounded domain of class* $C^{2+\nu}$, $\nu > 0$. *Suppose that the initial function* $\varrho_{0,\delta}$ *is positive, belongs to class* $C^{2+\nu}(\overline{\Omega})$, *and satisfies the compatibility condition* $\nabla_x \varrho_{0,\delta} \cdot \mathbf{n} = 0$ *on* $\partial\Omega$. *Furthermore, let*

$$\mathbf{u} \mapsto \varrho[\mathbf{u}]$$

be the solution mapping, which assigns to any $\mathbf{u} \in C([0,T]; C_0^2(\overline{\Omega}; R^N))$ *the unique solution* ϱ *of (7.19), the existence of which is guaranteed by Lemma 7.2.*

Then this mapping takes bounded sets in the space $C([0,T]; C_0^2(\overline{\Omega}, R^N))$ into bounded sets in the space

$$V \equiv \left\{ \begin{array}{l} \partial_t \varrho \in C([0,T]; C^\nu(\overline{\Omega})) \\ \varrho \in C([0,T]; C^{2+\nu}(\overline{\Omega})), \end{array} \right\}$$

and

$$\mathbf{u} \in C([0,T]; C_0^2(\overline{\Omega}, R^N)) \mapsto \varrho[\mathbf{u}] \in C^1([0,T] \times \overline{\Omega})$$

is continuous.

Remark In accordance with our agreement from Chapter 2, the symbol C_0^k stands for the subspace of C^k of functions for which all derivatives up to order k vanish on $\partial\Omega$.

Proof Boundedness of the mapping $\mathbf{u} \mapsto \varrho[\mathbf{u}]$ follows via iterating (bootstrapping) the estimates (7.20), (7.21). Moreover, let

$$\mathbf{u}_n \to \mathbf{u} \text{ in } C([0,T], C_0^2(\overline{\Omega}, R^N)).$$

Since the images $\varrho_n[\mathbf{u}_n]$ are bounded in V, one can use the Arzelà–Ascoli theorem (Theorem 2.1) to deduce that

$$\varrho_n \to \varrho \text{ in } C^1([0,T] \times \overline{\Omega})$$

at least for a suitable subsequence.

On the other hand, it is clear that the limit function ϱ is a weak solution of (7.19) in the sense of integral identity (7.23). Thus by virtue of Lemma 7.1, we have $\varrho = \varrho[\mathbf{u}]$, which completes the proof.

$\square$

7.3.2 The thermal energy equation

Pursuing the same strategy as in the preceding part, we consider the regularized thermal energy equation:

$$\left\{ \begin{array}{c} \partial_t \left((\delta + \varrho) Q(\vartheta) \right) + \mathrm{div}_x (\varrho Q(\vartheta) \mathbf{u}) - \Delta K(\vartheta) + \delta \vartheta^{\alpha+1} \\ = (1-\delta) \mathbb{S} : \nabla_x \mathbf{u} - \vartheta\, p_\vartheta\, \mathrm{div}_x \mathbf{u}, \\ \nabla_x \vartheta \cdot \mathbf{n}|_{\partial\Omega} = 0,\ \vartheta(0) = \vartheta_{0,\delta}, \end{array} \right\} \qquad (7.24)$$

where $\mathbf{u} \in C([0,T]; C_0^2(\overline{\Omega}; R^N))$, $\varrho[\mathbf{u}]$ are given functions corresponding to a strictly positive initial density distribution $\varrho_{0,\delta}$.

To begin with, we examine the question of *uniqueness* of strong solutions to problem (7.24).

Lemma 7.3 *Let $\mathbf{u} \in C([0,T]; C_0^2(\overline{\Omega}; R^N))$, $\varrho = \varrho[\mathbf{u}]$, and a non-negative function $\vartheta_{0,\delta} \in L^1(\Omega)$ be given.*

Then there exists at most one non-negative function ϑ,

$$\vartheta \in L^\infty((0,T) \times \Omega) \cap L^r(0,T;W^{1,r}(\Omega)),$$
$$\mathcal{K}(\vartheta) \in L^r(0,T;W^{2,r}(\Omega)),$$
$$\partial_t \vartheta \in L^r((0,T) \times \Omega) \quad \text{with } r > 1,$$

which solves the equation (7.24) a.a. on the set $(0,T) \times \Omega$, and satisfies the boundary and initial conditions in the sense of traces.

Proof   Let ϑ_1, ϑ_2 be two solutions of (7.24) with the same data. Accordingly, we have

$$(\delta + \varrho)\partial_t\Big(Q(\vartheta_1) - Q(\vartheta_2)\Big)$$

$$+ \varrho\mathbf{u} \cdot \nabla_x\Big(Q(\vartheta_1) - Q(\vartheta_2)\Big) - \Delta\Big(\mathcal{K}(\vartheta_1) - \mathcal{K}(\vartheta_2)\Big)$$

$$= (\vartheta_2 - \vartheta_1)p_\vartheta \operatorname{div}_x\mathbf{u} - \delta(\vartheta_1^{\alpha+1} - \vartheta_2^{\alpha+1}) - \varepsilon\Delta\varrho\Big(Q(\vartheta_1) - Q(\vartheta_2)\Big) \qquad (7.25)$$

a.a. on $(0,T) \times \Omega$,

$$\vartheta^1(0) - \vartheta^2(0) = \nabla_x(\mathcal{K}(\vartheta_1) - \mathcal{K}(\vartheta_2)) \cdot \mathbf{n}|_{\partial\Omega} = 0 \text{ in the sense of traces.}$$

The next step is to multiply (7.25) by the expression $\operatorname{sgn}(\vartheta_1 - \vartheta_2)$. By virtue of Theorem 2.1.11 in [127], one has $|Q(\vartheta_1) - Q(\vartheta_2)| \in W^{1,r}((0,T) \times \Omega)$ and, moreover,

$$\partial_t|Q(\vartheta_1) - Q(\vartheta_2)| = \operatorname{sgn}(\vartheta_1 - \vartheta_2)\partial_t\Big(Q(\vartheta_1) - Q(\vartheta_2)\Big),$$

$$\nabla_x|Q(\vartheta_1) - Q(\vartheta_2)| = \operatorname{sgn}(\vartheta_1 - \vartheta_2)\nabla_x\Big(Q(\vartheta_1) - Q(\vartheta_2)\Big).$$

Thus we obtain

$$\delta\partial_t|Q(\vartheta_1) - Q(\vartheta_2)| + \varrho\Big(\partial_t|Q(\vartheta_1) - Q(\vartheta_2)| + \mathbf{u} \cdot \nabla_x|Q(\vartheta_1) - Q(\vartheta_2)|\Big)$$

$$\leq \Delta\Big(\mathcal{K}(\vartheta_1) - \mathcal{K}(\vartheta_2)\Big)\operatorname{sgn}(\mathcal{K}(\vartheta_1) - \mathcal{K}(\vartheta_2)) + |\vartheta_1 - \vartheta_2||p_\vartheta \operatorname{div}_x\mathbf{u}|$$

$$+ \varepsilon|\Delta\varrho||Q(\vartheta_1) - Q(\vartheta_2)|. \qquad (7.26)$$

Integrating (7.26) yields

$$\int_\Omega (\delta + \varrho(\tau))|(Q(\vartheta_1) - Q(\vartheta_2))(\tau)| \, d\mathbf{x}$$

$$\leq \int_0^\tau \int_\Omega \Delta\Big(\mathcal{K}(\vartheta_1) - \mathcal{K}(\vartheta_2)\Big)\operatorname{sgn}(\mathcal{K}(\vartheta_1) - \mathcal{K}(\vartheta_2)) \, d\mathbf{x} \, dt$$

$$+ c\int_0^\tau \int_\Omega |Q(\vartheta_1) - Q(\vartheta_2)|\Big(|p_\vartheta\operatorname{div}_x \mathbf{u}| + 2\varepsilon|\Delta\varrho|\Big) \, d\mathbf{x} \, dt$$

for any $\tau \geq 0$. Note that, in accordance with hypothesis (3.5),

$$|\vartheta_1 - \vartheta_2| \leq c|Q(\vartheta_1) - Q(\vartheta_2)|.$$

Since ϱ is non-negative, this relation, together with Gronwall's lemma and estimate (7.22), yields the desired result $\vartheta_1 \equiv \vartheta_2$ as soon as we observe that

$$\int_0^\tau \int_\Omega \Delta\Big(\mathcal{K}(\vartheta_1) - \mathcal{K}(\vartheta_2)\Big)\mathrm{sgn}(\mathcal{K}(\vartheta_1) - \mathcal{K}(\vartheta_2)) \, \mathrm{dx} \, \mathrm{dt} \, \leq 0. \qquad (7.27)$$

Indeed for a smooth, non-decreasing, and odd function σ, and smooth functions ϑ_1, ϑ_2 satisfying the homogeneous Neumann boundary conditions, one has

$$\int_0^\tau \int_\Omega \Delta\Big(\mathcal{K}(\vartheta_1) - \mathcal{K}(\vartheta_2)\Big)\sigma(\mathcal{K}(\vartheta_1) - \mathcal{K}(\vartheta_2)) \, \mathrm{dx} \, \mathrm{dt}$$

$$= -\int_0^\tau \int_\Omega \sigma'(\vartheta_1 - \vartheta_2)\Big|\nabla_x(\mathcal{K}(\vartheta_1) - \mathcal{K}(\vartheta_2))\Big|^2 \, \mathrm{dx} \, \mathrm{dt} \leq 0.$$

Consequently, in order to obtain (7.27), it is enough to approximate the function sgn as well as ϑ_1, ϑ_2 by a sequence of smooth functions and use the Lebesgue convergence theorem.

$\square$

Focusing now on the existence problem, we first observe that (7.24) represents a *quasilinear parabolic equation* even if $\mathbf{u}$ and ϱ are fixed. Although the existence of solutions to this type of equation is well developed (see e.g. [55, 75]), it seems quite difficult to find a result that would apply directly to (7.24). The problem is low regularity of certain terms with respect to time. However, a straightforward modification of the standard approach yields the following result.

Lemma 7.4 *Let $\Omega \subset R^N$ be a bounded domain with boundary of class $C^{2+\nu}$. Let $\mathbf{u} \in C([0,T]; C_0^2(\overline{\Omega}))$, $\varrho = \varrho[\mathbf{u}]$ be given, with $\varrho_{0,\delta}$ strictly positive. Finally, assume that $\vartheta_{0,\delta} \in L^\infty(\Omega) \cap W^{1,2}(\Omega)$ is a non-negative function.*

Then the problem (7.24) possesses a (unique) strong solution ϑ which is non-negative and such that

$$\vartheta \in L^\infty((0,T) \times \Omega) \cap L^2(0,T; W^{1,2}(\Omega)),$$

$$\mathcal{K}(\vartheta) \in L^2(0,T; W^{2,2}(\Omega)),$$

$$\partial_t \vartheta \in L^2(0,T; L^2(\Omega)).$$

Moreover, we have

$$\|\partial_t \vartheta\|_{L^2((0,T)\times\Omega)} + \|\vartheta\|_{L^\infty((0,T)\times\Omega)} + \|\mathcal{K}(\vartheta)\|_{L^2(0,T;W^{2,2}(\Omega))}$$

$$\leq c\Big(\delta, \varepsilon, \|\mathbf{u}\|_{C([0,T];C_0^2(\overline{\Omega};R^N))}\Big). \qquad (7.28)$$

Proof (i) The informal idea of the proof is to regularize the equation with respect to t via the convolution operators defined in (2.3), (2.4), to apply the available existence theory of [75], and to get rid of the regularization through a limit process.

To this end, problem (7.24) is replaced by

$$\partial_t \Big((\delta + [\varrho]_t^\omega) Q(\vartheta) \Big) + \text{div}_x([\varrho\mathbf{u}]_t^\omega Q(\vartheta)) - \Delta \mathcal{K}(\vartheta) + \delta\vartheta^{\alpha+1}$$

$$= (1-\delta)[\mathbb{S}:\nabla_x\mathbf{u}]_t^\omega - \vartheta[p_\vartheta \ \text{div}_x\mathbf{u}]_t^\omega,$$

$$\nabla_x\vartheta \cdot \mathbf{n}|_{\partial\Omega} = 0, \ \vartheta(0) = \vartheta_{0,\delta}, \tag{7.29}$$

where $v \mapsto [v]_t^\omega$ denotes the convolution operator introduced in (2.3), (2.4).

As a matter of fact, the regularized quantities are defined only on a bounded time interval and must be extended to be zero outside $[0, T]$. More precisely,

$$[v]_t^\varepsilon(t, \mathbf{x}) \equiv \int_0^T \psi_\omega(t-s)v(s, \mathbf{x}) \ ds,$$

where

$$\left\{ \begin{array}{c} \psi_\omega \in \mathcal{D}(R), \ \text{supp}[\psi_\omega] \subset [-\omega, \omega], \\ \psi \text{ is even on } R, \text{ and non-increasing on } [0, \infty), \\ \int_R \psi_\omega(t) \ dt = 1. \end{array} \right\}$$

Problem (7.29) is a scalar quasilinear equation with differentiable (both in t and $\mathbf{x}$) coefficients. Note in this respect that the approximate densities $\varrho[\mathbf{u}]$ are positive bounded below away from zero on $[0, T] \times \overline{\Omega}$.

Accordingly, Theorem 8.1 in Chapter V of [75] can be applied to deduce that problem (7.29) possesses a unique classical solution ϑ for any $\omega > 0$.

(ii) The next step of the proof is to show uniform bounds on ϑ. These estimates are based on the *maximum principle* or *comparison theorem* for scalar parabolic equations. To this end, equation (7.29) can be written as

$$([\varrho]_t^\omega + \delta)Q'(\vartheta)\partial_t\vartheta + Q'(\vartheta)[\varrho\mathbf{u}]_t^\omega \cdot \nabla_x\vartheta - \text{div}_x(\kappa(\vartheta)\nabla_x\vartheta) + \delta\vartheta^{\alpha+1}$$

$$= (1-\delta)[\mathbb{S}:\nabla_x\mathbf{u}]_t^\omega - \vartheta[p_\vartheta \ \text{div}_x \ \mathbf{u}]_t^\omega + \varepsilon Q(\vartheta)[\Delta\varrho]_t^\omega, \tag{7.30}$$

which can be viewed as a *linear* scalar parabolic equation for ϑ with time dependent coefficients.

As the term $[\mathbb{S}:\nabla_x u]_t^\omega$ is always non-negative, the function $\vartheta = 0$ is a subsolution for (7.30) and, consequently,

$$\vartheta(t, \mathbf{x}) \geq 0 \quad \text{for all } t \in [0, T], \mathbf{x} \in \Omega \tag{7.31}$$

(see e.g. Chapter 3 of [106]).

On the other hand, it is easy to observe that

$$\delta\vartheta^{\alpha+1} > (1-\delta)[\mathbb{S}:\nabla_x\mathbf{u}]_t^\omega - \vartheta[p_\vartheta \ \text{div}_x\mathbf{u}]_t^\omega + \varepsilon[\Delta\varrho]_t^\omega Q(\vartheta)$$

for all $\vartheta > 0$ sufficiently large provided $\delta > 0$. Accordingly, since the initial data $\vartheta_{0,\delta}$ are bounded, we can find $\overline{\vartheta} > 0$ large enough so that the constant function

$\vartheta = \overline{\vartheta}$ is a supersolution for (7.30); and we get

$$\vartheta(t, \mathbf{x}) \le \overline{\vartheta} \quad \text{for all } t \in [0, T], \ \mathbf{x} \in \Omega. \tag{7.32}$$

Of course, the bound in (7.32) depends on δ, ε as well as on $\mathbf{u}$.

(iii) Now we establish *energy estimates* resulting from the multiplication of (7.29) on ϑ:

$$\int_\Omega \left(\delta + [\varrho]_t^\omega\right)\left(Q(\vartheta)\vartheta - \tilde{Q}(\vartheta)\right)(T)\,\mathrm{d}\mathbf{x} + \varepsilon\int_0^T \int_\Omega \Delta[\varrho]_t^\omega \tilde{Q}(\vartheta)\,\mathrm{d}\mathbf{x}\,\mathrm{d}t$$

$$+ \int_0^T \int_\Omega \kappa(\vartheta)|\nabla_x\vartheta|^2 + \delta\vartheta^{\alpha+2}\,\mathrm{d}\mathbf{x}\,\mathrm{d}t$$

$$= \int_\Omega \left(\delta + [\varrho]_t^\omega\right)\left(Q(\vartheta)\vartheta - \tilde{Q}(\vartheta)\right)(0)\,\mathrm{d}\mathbf{x}$$

$$+ \int_0^T \int_\Omega (1-\delta)[\mathbb{S} : \nabla_x\mathbf{u}]_t^\omega \vartheta - \vartheta^2[p_\vartheta\,\mathrm{div}_x\mathbf{u}]_t^\omega\,\mathrm{d}\mathbf{x}\,\mathrm{d}t,$$

where

$$\tilde{Q}(\vartheta) \equiv \int_0^\vartheta Q(z)\,\mathrm{d}z.$$

In view of the uniform estimates (7.31), (7.32) we deduce that

$$\vartheta \text{ is bounded in } L^2(0, T; W^{1,2}(\Omega)) \tag{7.33}$$

independently of ω.

(iv) Next we multiply (7.29) by $\kappa(\vartheta)\partial_t\vartheta$ to obtain

$$\frac{\mathrm{d}}{\mathrm{d}t}\int_\Omega \frac{1}{2}|\nabla_x\mathcal{K}(\vartheta)|^2\,\mathrm{d}\mathbf{x} + \int_\Omega (\delta + [\varrho]_t^\omega)Q'(\vartheta)\kappa(\vartheta)|\partial_t\vartheta|^2\,\mathrm{d}\mathbf{x}$$

$$+ \varepsilon\int_\Omega \Delta[\varrho]_t^\omega Q(\vartheta)\kappa(\vartheta)\partial_t\vartheta\,\mathrm{d}\mathbf{x} + \delta\int_\Omega \vartheta^{\alpha+1}\kappa(\vartheta)\partial_t\vartheta\,\mathrm{d}\mathbf{x}$$

$$= \int_\Omega (1-\delta)[\mathbb{S} : \nabla_x\mathbf{u}]_t^\omega \kappa(\vartheta)\partial_t\vartheta - \kappa(\vartheta)\vartheta\,\partial_t\vartheta\,[p_\vartheta\,\mathrm{div}_x\mathbf{u}]_t^\omega\,\mathrm{d}\mathbf{x}. \tag{7.34}$$

In view of the estimates we have already obtained, relation (7.34) yields a bound

$$\partial_t\vartheta \text{ bounded in } L^2((0, T) \times \Omega). \tag{7.35}$$

To conclude, it is a routine matter to express $\Delta\mathcal{K}(\vartheta)$ in (7.30) to obtain

$$\Delta\mathcal{K}(\vartheta) \text{ bounded in } L^2((0, T) \times \Omega),$$

which implies a bound on $\mathcal{K}(\vartheta)$ in $L^2(0, T; W^{2,2}(\Omega))$ independent of ω (see e.g. [2]). This completes the proof of (7.28).

(v) Finally, given (7.28) it is easy to pass to the limit for $\omega \to 0$ in (7.29). Denoting by $\{\vartheta_\omega\}_{\omega>0}$ the corresponding (classical) solutions, we have

$$\vartheta_\omega \to \vartheta \text{ weakly-* in } L^\infty((0,T) \times \Omega),$$

$$\mathcal{K}(\vartheta_\omega) \to \mathcal{K}(\vartheta) \text{ weakly in } L^2(0,T;W^{2,2}(\Omega)),$$

$$\partial_t \vartheta_\omega \to \partial_t \vartheta \text{ weakly in } L^2((0,T) \times \Omega),$$

where, as is easily seen from the compactness of the imbedding $W^{1,2}((0,T) \times \Omega) \subset L^2((0,T) \times \Omega)$, the limit function ϑ solves (7.24).

$\square$

Similarly to Proposition 7.1, the basic properties of the solution operator $\mathbf{u} \mapsto \vartheta[\mathbf{u}]$ related to problem (7.24) can be summarized as follows.

Proposition 7.2 *Let $\Omega \subset R^N$ be a bounded domain of class $C^{2+\nu}$, $\nu > 0$. Assume that $\vartheta_{0,\delta}$ is a non-negative function belonging to the space $L^\infty \cap W^{1,2}(\Omega)$. Furthermore, suppose that $\mathbf{u} \in C([0,T]; C_0^2(\overline{\Omega}, R^N))$ is a given velocity field, and $\varrho = \varrho[\mathbf{u}]$ is the solution of problem (7.19) emanating from a strictly positive initial function $\varrho_{0,\delta}$. Denote*

$$\mathbf{u} \mapsto \vartheta[\mathbf{u}]$$

the solution operator, which assigns to $\mathbf{u} \in C([0,T]; C_0^2(\overline{\Omega}, R^N))$ the unique solution ϑ of (7.24) with $\varrho = \varrho[\mathbf{u}]$.

Then $\mathbf{u} \mapsto \vartheta[\mathbf{u}]$ maps bounded sets in $C([0,T]; C_0^2(\overline{\Omega}, R^N))$ into bounded sets in the set

$$Y \equiv \left\{ \vartheta \in L^\infty((0,T) \times \Omega) \cap L^2(0,T;W^{1,2}(\Omega)), \quad \begin{array}{c} \partial_t \vartheta \in L^2((0,T) \times \Omega), \\ \mathcal{K}(\vartheta) \in L^2(0,T;W^{2,2}(\Omega)), \end{array} \right\}$$

and the mapping

$$\mathbf{u} \in C([0,T]; C_0^2(\overline{\Omega}, R^N)) \mapsto \vartheta \in L^2(0,T;W^{1,2}(\Omega))$$

is continuous.

Remark By virtue of the same arguments as in Proposition 7.1, boundedness of the operator $\mathbf{u} \mapsto \vartheta[\mathbf{u}]$ is a consequence of *a priori* estimates while continuity follows from the uniqueness of the solutions established above. Note that $\partial_t \mathcal{K}(\vartheta)$ is bounded in $L^2((0,T) \times \Omega)$ which yields $\mathcal{K}(\vartheta)$ (and ϑ) compact in $L^2(0,T;W^{1,2}(\Omega))$.

7.3.3 *Approximate solutions: local existence*

Having collected all preliminary results we turn our attention to the approximate problem represented by the integral equation (7.18). Following the standard approach we solve (7.18) on a possibly short time interval via a fixed point argument, establish estimates of the solution that are independent of the length of this interval, and iterate this procedure to obtain, after a finite number of steps, a solution defined on the whole time interval $[0, T]$.

In order to carry out this idea, we must set up some technical apparatus. Consider a family of linear operators

$$\mathcal{M}[\varrho] : X_n \to X_n^*, \quad \langle \mathcal{M}[\varrho]\mathbf{v}, \mathbf{w} \rangle \equiv \int_\Omega \varrho\, \mathbf{v} \cdot \mathbf{w} \, \mathrm{d}x.$$

Here X_n is considered as a (finite dimensional) Hilbert space with a scalar product induced by the standard L^2-norm. As always, the symbol X_n^* stands for the dual space of X_n.

It is easy to see that the operator $\mathcal{M}$ is invertible provided ϱ is strictly positive on Ω, and

$$\| \mathcal{M}^{-1}[\varrho] \|_{\mathcal{L}(X_n^*; X_n)} \leq \frac{1}{\inf_\Omega \varrho}.$$

Moreover, the identity

$$\mathcal{M}^{-1}[\varrho^1] - \mathcal{M}^{-1}[\varrho^2] = \mathcal{M}^{-1}[\varrho^2]\Big(\mathcal{M}[\varrho^2] - \mathcal{M}[\varrho^1]\Big)\mathcal{M}^{-1}[\varrho^1]$$

can be used to obtain

$$\| \mathcal{M}^{-1}[\varrho^1] - \mathcal{M}^{-1}[\varrho^2] \|_{\mathcal{L}(X_n^*; X_n)} \leq c(n, \underline{\varrho})\|\varrho_1 - \varrho_2\|_{L^1(\Omega)} \tag{7.36}$$

for any ϱ_1, ϱ_2 such that

$$\inf_\Omega \varrho^1, \quad \inf_\Omega \varrho^2 \geq \underline{\varrho}.$$

Let us remark that the C and L^1-norm are *equivalent* on the finite dimensional space X_n.

Now the integral equation (7.18) can be rephrased as

$$\mathbf{u}_n(t) = \mathcal{M}^{-1}[\varrho(t)]\Big(\mathbf{m}_{0,\delta}^* + \int_0^t \mathcal{N}[\mathbf{u}_n(s), \varrho(s), \vartheta(s)] \, \mathrm{d}s\Big), \tag{7.37}$$

with

$$\mathbf{m}_{0,\delta}^* \in X_n^*, \quad \langle \mathbf{m}_{0,\delta}^*, \eta \rangle \equiv \int_\Omega \mathbf{m}_{0,\delta} \cdot \eta \, \mathrm{d}x \quad \text{for any } \eta \in X_n,$$

and

$$\mathcal{N} : X_n \to X_n^*,$$

$$\langle \mathcal{N}[\mathbf{u}_n, \varrho, \vartheta], \eta \rangle \equiv \int_\Omega [\varrho \mathbf{u}_n \otimes \mathbf{u}_n - \mathbb{S}_n] : \nabla_x \eta + [p(\varrho, \vartheta) + \delta \varrho^\beta]\mathrm{div}_x \, \eta \, \mathrm{d}x$$

$$+ \int_\Omega [\varrho \mathbf{f} - \varepsilon \nabla_x \mathbf{u}_n \nabla_x \varrho] \cdot \eta \, \mathrm{d}x,$$

where $\varrho = \varrho[\mathbf{u}_n]$ and $\vartheta = \vartheta[\mathbf{u}_n]$ are uniquely determined by $\mathbf{u}_n$.

Consider a bounded ball $\mathcal{B}$ in the space $C([0, T]; X_n)$,

$$\mathcal{B} \equiv \left\{ \mathbf{v} \in C([0, T]; X_n) \;\middle|\; \sup_{t \in [0,T]} \|\mathbf{v}(t) - \mathbf{u}_{0,\delta,n}\|_{X_n} \leq 1 \right\}$$

where the value of $\mathbf{u}_{0,\delta,n} \in X_n$ is uniquely determined through

$$\int_\Omega \varrho_{0,\delta} \mathbf{u}_{0,\delta,n} \cdot \eta \, \mathrm{d}\mathbf{x} = \int_\Omega \mathbf{m}_{0,\delta} \cdot \eta \, \mathrm{d}\mathbf{x} \quad \text{for all } \eta \in X_n.$$

Finally, define a mapping

$$\mathcal{T} : \mathcal{B} \to C([0, T]; X_n),$$

$$\mathcal{T}[\mathbf{u}] \equiv \mathcal{M}^{-1}[\varrho(t)] \left(\mathbf{m}_{0,\delta}^* + \int_0^t \mathcal{N}[\mathbf{u}(s), \varrho(s), \vartheta(s)] \, \mathrm{d}s \right).$$

By virtue of (7.36) and the estimates established in Propositions 7.1, 7.2, $\mathcal{T}$ maps the ball $\mathcal{B}$ into itself provided $T = T(n)$ is small enough. Moreover, by the same token, $\mathcal{T}$ is a continuous mapping and the image of $\mathcal{B}$ under $\mathcal{T}$ is a compact subset of $\mathcal{B}$. Consequently, one can use the Schauder theorem (see e.g. Theorem 2.1.2 in [101]) to conclude that there exists at least one fixed point $\mathbf{u}_n$, $\mathcal{T}[\mathbf{u}_n] = \mathbf{u}_n$—a solution of the integral equation (7.18) on $[0, T(n)]$.

Now, this procedure can be iterated as many times as necessary to reach $T(n) = T$ as long as there is a bound on $\mathbf{u}_n$ independent of $T(n)$. The existence of such a bound will follow from estimates derived in the next section.

7.3.4 *Approximate solutions: time-independent estimates*

In this section, we shall show that the *a priori* estimates obtained in Chapter 3 for exact solutions are compatible with our approximation scheme. Two types of bounds will be obtained: (i) estimates that are independent of time but may depend on the dimension n of X_n; (ii) estimates independent of n.

We start with the *energy estimates* discussed in Section 3.3. It follows from (7.18) that $\mathbf{u}_n$ is continuously differentiable and, consequently, the integral identity

$$\int_\Omega \partial_t(\varrho_n \mathbf{u}_n) \cdot \eta \, \mathrm{d}\mathbf{x}$$

$$= \int_\Omega [\varrho_n \mathbf{u}_n \otimes \mathbf{u}_n - \mathbb{S}_n] : \nabla_x \eta + [p(\varrho_n, \vartheta_n) + \delta\varrho_n^\beta] \operatorname{div}_x \eta \, \mathrm{d}\mathbf{x}$$

$$+ \int_\Omega [\varrho_n \mathbf{f} - \varepsilon \nabla_x \mathbf{u}_n \nabla_x \varrho_n] \cdot \eta \, \mathrm{d}\mathbf{x} \tag{7.38}$$

holds on $(0, T(n))$ for any $\eta \in X_n$. Moreover,

$$\int_\Omega \varrho_{0,\delta} \mathbf{u}_{0,\delta,n}(0) \cdot \eta \, \mathrm{d}\mathbf{x} = \int_\Omega \mathbf{m}_{0,\delta} \cdot \eta \, \mathrm{d}\mathbf{x} \quad \text{for any } \eta \in X_n. \tag{7.39}$$

In accordance with our convention, we write $\varrho_n = \varrho[\mathbf{u}_n]$, $\vartheta_n = \vartheta[\mathbf{u}_n]$.

By virtue of hypothesis (3.1), the elastic pressure component p_e can be written in the form

$$p_e(\varrho) = p_m(\varrho) + p_b(\varrho), \quad |p_b| \leq M,$$

where p_m is a non-decreasing function of ϱ. We set

$$P_m(\varrho) \equiv \int_1^\varrho \frac{p_m(z)}{z^2} \, dz.$$

Taking $\eta = \mathbf{u}_n(t)$ in (7.38) and integrating by parts, we obtain

$$\frac{d}{dt} \int_\Omega \varrho_n \left(\frac{1}{2} |\mathbf{u}_n|^2 + P_m(\varrho_n) + \frac{\delta}{\beta - 1} \varrho_n^{\beta - 1} \right) dx$$

$$+ \frac{\varepsilon}{2} \int_\Omega \Delta \varrho_n |\mathbf{u}_n|^2 \, dx + \varepsilon \int_\Omega |\nabla_x \varrho_n|^2 \left(\frac{p_m'(\varrho_n)}{\varrho_n} + \delta \beta \varrho_n^{\beta - 2} \right) dx$$

$$= \int_\Omega p_b(\varrho_n) \, \text{div}_x \, \mathbf{u}_n \, dx$$

$$+ \int_\Omega \vartheta_n p_\vartheta(\varrho_n) \, \text{div}_x \mathbf{u}_n - \mathbb{S}_n : \nabla_x \mathbf{u}_n + \varrho_n \mathbf{f} \cdot \mathbf{u}_n - \varepsilon(\nabla_x \mathbf{u}_n \nabla_x \varrho) \cdot \mathbf{u}_n \, dx.$$

$$(7.40)$$

Let us recall that

$$p_m(\varrho_n) \, \text{div}_x \mathbf{u}_n = -\text{div}_x \left(\varrho_n P_m(\varrho_n) \mathbf{u}_n \right) - \partial_t \left(\varrho_n P_m(\varrho_n) \right)$$

$$+ \varepsilon \Delta \varrho_n \left(P_m(\varrho_n) + \frac{p_m(\varrho_n)}{\varrho_n} \right),$$

where

$$\int_\Omega \Delta \varrho_n \left(P_m(\varrho_n) + \frac{p_m(\varrho_n)}{\varrho_n} \right) dx = -\int_\Omega \frac{p_m'(\varrho_n)}{\varrho_n} |\nabla_x \varrho_n|^2 \, dx;$$

and

$$\int_\Omega \partial_t(\varrho_n \mathbf{u}_n) \cdot \mathbf{u}_n - (\varrho_n \mathbf{u}_n \otimes \mathbf{u}_n) : \nabla_x \mathbf{u}_n \, dx$$

$$= \frac{d}{dt} \int_\Omega \frac{1}{2} \varrho_n |\mathbf{u}_n|^2 \, dx + \frac{1}{2} \int_\Omega \left(\partial_t \varrho_n + \text{div}_x(\varrho_n \mathbf{u}_n) \right) |\mathbf{u}_n|^2 \, dx$$

$$= \frac{d}{dt} \int_\Omega \frac{1}{2} \varrho_n |\mathbf{u}_n|^2 \, dx + \frac{\varepsilon}{2} \int_\Omega \Delta \varrho_n |\mathbf{u}_n|^2 \, dx.$$

Now, equation (7.24) integrated over Ω and added to (7.40) yields

$$\frac{\mathrm{d}}{\mathrm{d}t} \int_\Omega \varrho_n \Big(\frac{1}{2}|\mathbf{u}_n|^2 + P_\mathrm{m}(\varrho_n) + \frac{\delta}{\beta-1}\varrho_n^{\beta-1} + Q(\vartheta_n)\Big) + \delta Q(\vartheta_n)\,\mathrm{dx}$$

$$+ \delta \int_\Omega \mathbb{S}_n : \nabla_x \mathbf{u}_n\,\mathrm{dx} + \int_\Omega \delta\vartheta_n^{\alpha+1} + \varepsilon|\nabla_x\varrho_n|^2\Big(\frac{p'_\mathrm{m}(\varrho_n)}{\varrho_n} + \delta\beta\varrho_n^{\beta-2}\Big)\,\mathrm{dx}$$

$$= \int_\Omega p_b(\varrho_n)\,\mathrm{div}_x\mathbf{u}_n + \varrho_n\mathbf{f}\cdot\mathbf{u}_n\,\mathrm{dx}. \tag{7.41}$$

Note that we got rid of the integral $\varepsilon\int_\Omega \Delta\varrho_n|\mathbf{u}_n|^2\,\mathrm{dx}$ thanks to the extra term $\varepsilon\nabla_x\mathbf{u}_n\nabla_x\varrho_n$ in (7.6).

Moreover, one can integrate (7.19) to recover the principle of total mass conservation:

$$\int_\Omega \varrho(t)\,\mathrm{dx} = \int_\Omega \varrho_{0,\delta}\,\mathrm{dx} \quad \text{for any } t \geq 0. \tag{7.42}$$

Additionaly, integration of (7.41) in time reveals a modified *energy equality*

$$\int_\Omega \varrho_n(\tau)\Big(\frac{1}{2}|\mathbf{u}_n|^2 + P_\mathrm{m}(\varrho_n) + \frac{\delta}{\beta-1}\varrho_n^{\beta-1} + Q(\vartheta_n)\Big)(\tau) + \delta Q(\vartheta_n)(\tau)\,\mathrm{dx}$$

$$+ \delta \int_0^\tau \int_\Omega \mathbb{S}_n : \nabla_x\mathbf{u}_n + \vartheta_n^{\alpha+1}\,\mathrm{dx}\,\mathrm{dt}$$

$$+ \varepsilon \int_0^\tau \int_\Omega |\nabla_x\varrho_n|^2\Big(\frac{p'_\mathrm{m}(\varrho_n)}{\varrho_n} + \delta\beta\varrho_n^{\beta-2}\Big)\,\mathrm{dx}\,\mathrm{dt}$$

$$= \int_\Omega \frac{1}{2}\mathbf{m}_{0,\delta}\cdot\mathbf{u}_n(0) + \varrho_{0,\delta}P_\mathrm{m}(\varrho_{0,\delta}) + \frac{\delta}{\beta-1}\varrho_{0,\delta}^\beta + (\delta+\varrho_{0,\delta})Q(\vartheta_{0,\delta})\,\mathrm{dx}$$

$$+ \int_0^\tau \int_\Omega p_b(\varrho_n)\,\mathrm{div}_x\mathbf{u}_n + \varrho_n\mathbf{f}\cdot\mathbf{u}_n\,\mathrm{dx}\,\mathrm{dt} \quad \text{for any } \tau \in [0, T(n)]. \tag{7.43}$$

By virtue of (7.15), (7.42), we have

$$\int_\Omega \mathbf{m}_{0,\delta}\cdot\mathbf{u}_{0,\delta,n}(0)\,\mathrm{dx} \leq \frac{1}{2}\int_\Omega \frac{|\mathbf{m}_{0,\delta}|^2}{\varrho_{0,\delta}} + \varrho_{0,\delta}|\mathbf{u}_{0,\delta,n}|^2\,\mathrm{dx}$$

$$= \frac{1}{2}\int_\Omega \frac{|\mathbf{m}_0|^2}{\varrho_{0,\delta}} + \mathbf{m}_{0,\delta}\cdot\mathbf{u}_{0,\delta,n}(0)\,\mathrm{dx},$$

$$\int_\Omega \varrho_{0,\delta}P_\mathrm{m}(\varrho_{0,\delta})\,\mathrm{dx} \leq \int_\Omega \varrho_{0,\delta}P_\mathrm{e}(\varrho_{0,\delta}) + c(1+\varrho_{0,\delta})\,\mathrm{dx},$$

and

$$\int_\Omega \varrho_n \mathbf{f} \cdot \mathbf{u}_n \, \mathrm{d}\mathbf{x} \leq \frac{1}{2} \mathrm{ess\,sup}_{t,\mathbf{x}} |\mathbf{f}(t,\mathbf{x})| \int_\Omega \varrho_n + \varrho_n |\mathbf{u}_n|^2 \, \mathrm{d}\mathbf{x}$$

$$= \frac{1}{2} \mathrm{ess\,sup}_{t,\mathbf{x}} |\mathbf{f}(t,\mathbf{x})| \Big(\int_\Omega \varrho_{0,\delta} \, \mathrm{d}\mathbf{x} + \int_\Omega \varrho_n |\mathbf{u}_n|^2 \, \mathrm{d}\mathbf{x} \Big);$$

whence the first term on the right-hand side of (7.43) is bounded in view of (7.17) while the second one may be treated with help of Gronwall's lemma to deduce that

$$\mathbf{u}_n \text{ is bounded in } L^2(0, T(n); W_0^{1,2}(\Omega; R^N))$$

by a constant that is independent of n and $T(n) \leq T$. Since all norms are equivalent on X_n, this implies that the approximate velocity fields $\mathbf{u}_n$ are bounded in $L^1(0, T(n); W^{1,\infty}(\Omega; R^N))$, in particular, by virtue of (7.22), the density ϱ_n is bounded both from below and from above by a constant independent of $T(n) \leq T$.

Since ϱ_n is bounded from below, one can use (7.43) to deduce uniform boundedness in t of $\mathbf{u}_n$ in the space $L^2(\Omega; R^N)$. Consequently, the functions $\mathbf{u}_n(t)$ remain bounded in X_n for any t independently of $T(n) \leq T$.

This extra bit of information concerning the boundedness of $\mathbf{u}_n$ together with Proposition 7.2 can be used to conclude that $\vartheta_n(t)$ is bounded in $W^{1,2}(\Omega)$. Here, we utilized the imbedding of the space Y defined in Proposition 7.2 into $C([0,T]; W^{1,2}(\Omega))$.

Thus we are allowed to iterate the previous local existence result to construct a solution defined on the whole time interval $[0,T]$.

Proposition 7.3 *For any fixed n and T, there exist functions*

$$\left\{ \begin{array}{c} \varrho_n \in C([0,T]; C^{2+\nu}(\overline{\Omega})), \quad \partial_t \varrho_n \in C([0,T]; C^\nu(\overline{\Omega})), \\[2mm] \mathbf{u}_n \in C^1([0,T]; X_n), \\[2mm] \vartheta_n \in L^\infty((0,T) \times \Omega) \cap L^2(0,T; W^{1,2}(\Omega)), \quad \mathcal{K}(\vartheta_n) \in L^2(0,T; W^{2,2}(\Omega)), \\[2mm] \partial_t \vartheta \in L^2((0,T) \times \Omega) \end{array} \right\}$$

solving problem (7.18) on the time interval $[0,T]$.

7.3.5 *Estimates independent of n*

Our goal now is to identify a limit for $n \to \infty$ of the approximate solutions ϱ_n, $\mathbf{u}_n$, ϑ_n as a solution of problem (7.3)–(7.11). In order to achieve this, additional estimates are needed.

To begin with, it is easy to see that the energy equation (7.43) yields

$$\sqrt{\varepsilon}\delta\varrho_n^{\beta/2} \text{ bounded in } L^2(0,T;W^{1,2}(\Omega)).$$

Evoking the imbedding $W^{1,2}(\Omega) \subset L^{2^*}(\Omega)$ (see Theorem 2.7) we get

$$\left\{\begin{array}{c} \varrho_n^{\beta} \text{ bounded in } L^1(0,T;L^{N/(N-2)}(\Omega)) \text{ if } N \geq 3, \\[2mm] \varrho_n^{\beta} \text{ bounded in } L^1(0,T;L^q(\Omega)), \ q > 1 \text{ arbitrary finite if } N = 2, \end{array}\right\}$$

where these bounds depend only on δ. Moreover, we have

$$\varrho_n^{\beta} \text{ is bounded in } L^{\infty}(0,T;L^1(\Omega)).$$

Consequently, the interpolation inequality (2.2) can be used to obtain

$$\|\varrho\|_{L^{\beta+1}((0,T)\times\Omega)} \leq c(\varepsilon,\delta), \ c(\varepsilon,\delta) \text{ independent of } n \tag{7.44}$$

provided

$$\beta > \frac{N}{2}. \tag{7.45}$$

Indeed we have

$$\int_{\Omega} \varrho^{\beta+1} \, \mathrm{d}\mathbf{x} = \|\varrho^{\beta}\|_{L^{(\beta+1)/\beta}(\Omega)}^{(\beta+1)/\beta} \leq \|\varrho^{\beta}\|_{L^{N/N-2}(\Omega)}^{N/2\beta} \|\varrho^{\beta}\|_{L^1(\Omega)}^{(2\beta+2-N)/2\beta} \text{ if } N = 3,$$

where we need $N/2\beta \leq 1$. The case $N = 2$ is similar.

Next, equation (7.19) multiplied on ϱ_n yields

$$\varepsilon \int_0^T \int_{\Omega} |\nabla_x \varrho_n|^2 \, \mathrm{d}\mathbf{x}\,\mathrm{d}t$$

$$\leq c(T)\left(\|\varrho_{0,\delta}\|_{L^2(\Omega)}^2 + \|\varrho_n\|_{L^{\infty}(0,T;L^4(\Omega))}^2\right)\left(\int_0^T \int_{\Omega} |\nabla_x \varrho_n|^2 \, \mathrm{d}\mathbf{x}\,\mathrm{d}t\right)^{1/2}.$$

Consequently, we deduce the estimate

$$\sqrt{\varepsilon}\|\nabla_x \varrho_n\|_{L^2((0,T)\times\Omega)} \leq c(\varepsilon,\delta) \tag{7.46}$$

provided $\beta \geq 4$.

Let us focus now on the modified thermal energy equation (7.24). Similarly to what was done in Chapter 4, we mutliply (7.24) by $h(\vartheta)$, where h enjoys the properties formulated in (4.48). Accordingly, we obtain

$$\partial_t\Big((\delta + \varrho_n)Q_h(\vartheta_n)\Big) + \mathrm{div}_x(\varrho_n Q_h(\vartheta_n)\mathbf{u}_n) - \Delta \mathcal{K}_h(\vartheta) + \delta \vartheta_n^{\alpha+1} h(\vartheta_n)$$

$$= (1 - \delta)h(\vartheta_n)\, \mathbb{S}_n : \nabla_x \mathbf{u}_n - \kappa(\vartheta_n)h'(\vartheta_n)|\nabla_x \vartheta_n|^2 - \vartheta_n h(\vartheta_n)p_\vartheta(\varrho_n)\, \mathrm{div}_x\, \mathbf{u}_n$$

$$+ \varepsilon\Delta\varrho_n\Big(Q_h(\vartheta_n) - Q(\vartheta_n)h(\vartheta_n)\Big), \tag{7.47}$$

where Q_h, $\mathcal{K}_h$ are determined by (4.49).

Integrating (7.47) over Ω yields

$$\frac{\mathrm{d}}{\mathrm{d}t}\int_\Omega (\delta + \varrho_n)Q_h(\vartheta_n)\,\mathrm{d}\mathbf{x} + \delta\int_\Omega \vartheta_n^{\alpha+1}h(\vartheta_n)\,\mathrm{d}\mathbf{x}$$

$$= \int_\Omega (1 - \delta)h(\vartheta_n)\mathbb{S}_n : \nabla_x \mathbf{u}_n - \kappa(\vartheta_n)h'(\vartheta_n)|\nabla_x\vartheta_n|^2\,\mathrm{d}\mathbf{x}$$

$$+ \int_\Omega \varepsilon(\nabla_x\varrho_n \cdot \nabla_x\vartheta_n)Q(\vartheta_n)h'(\vartheta_n) - \vartheta_n h(\vartheta_n)\, p_\vartheta(\varrho_n)\, \mathrm{div}_x\mathbf{u}_n\,\mathrm{d}\mathbf{x}. \tag{7.48}$$

In particular, the choice $h(\vartheta) = (1 + \vartheta)^{-1}$ leads to relations

$$-\int_\Omega \kappa(\vartheta_n)h'(\vartheta_n)|\nabla_x\vartheta_n|^2\,\mathrm{d}\mathbf{x} \geq c\int_\Omega |\nabla_x\vartheta_n^{\alpha/2}|^2\,\mathrm{d}\mathbf{x}$$

while

$$\varepsilon\Big|\int_\Omega \nabla_x\varrho_n \cdot \nabla_x\vartheta_n Q(\vartheta_n)h'(\vartheta_n)\,\mathrm{d}\mathbf{x}\Big| \leq \varepsilon\big\|\nabla_x\varrho_n\big\|_{L^2(\Omega)}\big\|\nabla_x\vartheta_n^{\alpha-1/2}\big\|_{L^2(\Omega)},$$

and

$$\Big|\int_\Omega \vartheta_n h(\vartheta_n)p_\vartheta(\varrho_n)\mathrm{div}_x\mathbf{u}_n\,\mathrm{d}\mathbf{x}\Big| \leq c\|p_\vartheta(\varrho_n)\|_{L^2(\Omega)}\|\mathrm{div}_x\mathbf{u}_n\|_{L^2(\Omega)},$$

where we have used hypotheses (3.3)–(3.6). It follows from hypothesis (3.2) and the energy estimates (7.43) that the right-hand side of the last inequality is bounded in $L^1(0,T)$ by a constant that depends only on δ.

Consequently, (7.48) integrated with respect to t together with the energy estimates (7.43) yield a bound

$$\|\nabla_x\vartheta_n^{\alpha/2}\|_{L^2((0,T)\times\Omega)} \leq c(\delta)$$

which is independent of n.

On the point of conclusion, we put together all estimates obtained in this section.

Proposition 7.4   *Assume that $\beta > \max\{4, \frac{N}{2}\}$.*

Then the approximate solutions constructed in Proposition 7.3 satisfy the following estimates:

$$\left. \begin{aligned} \|\varrho_n\|_{L^\infty(0,T;L^\gamma(\Omega))} \\[2mm] \|\varrho_n\|_{L^\infty(0,T;L^\beta(\Omega))} \\[2mm] \sqrt{\varepsilon}\|\nabla_x\varrho_n\|_{L^2((0,T)\times\Omega)} \end{aligned} \right\} \le c(\delta); \tag{7.49}$$

$$\|\varrho_n\|_{L^{\beta+1}((0,T)\times\Omega)} \le c(\varepsilon,\delta); \tag{7.50}$$

$$\|\mathbf{u}_n\|_{L^2(0,T;W_0^{1,2}(\Omega;R^N))} \le c(\delta); \tag{7.51}$$

$$\left. \begin{aligned} \|\vartheta_n\|_{L^{\alpha+1}((0,T)\times\Omega)}, \\[2mm] \|\nabla_x\vartheta^{\alpha/2}\|_{L^2((0,T)\times\Omega)} \end{aligned} \right\} \le c(\delta) \tag{7.52}$$

and

$$\|\sqrt{\varrho_n}\mathbf{u}_n\|_{L^\infty(0,T;L^2(\Omega;R^N))} \le c(\delta); \tag{7.53}$$

$$\|\varrho_n Q(\vartheta_n)\|_{L^\infty(0,T;L^1(\Omega))} \le c(\delta), \tag{7.54}$$

where all constants are independent of n.

7.3.6   The first level approximate solutions

At this stage we are ready to pass to the limit for $n \to \infty$ in the sequence of *approximate solutions* $\{\varrho_n, \mathbf{u}_n, \vartheta_n\}_{n=1}^\infty$ in order to obtain a solution to the system (7.3)–(7.11). To this end, we assume henceforth that the system of test functions $\{\eta_j\}_{j=1}^\infty$ forms a dense set, say, in the space $C_0^1(\overline{\Omega}; R^N)$.

It follows from equation (7.19) and the estimates obtained in Proposition 7.4 that the time derivative $\partial_t\varrho_n$ is bounded in the space $L^2(0,T;W^{-1,2}(\Omega))$ provided $\beta \ge N$. Consequently, one can use the Aubin–Lions lemma (see Theorem V.1 in Chapter 1 of [81]) to deduce that the sequence $\{\varrho_n\}_{n=1}^\infty$ contains a subsequence (not relabeled) such that

$$\varrho_n \to \varrho \text{ in } L^\beta((0,T) \times \Omega), \tag{7.55}$$

where ϱ is a non-negative function.

Moreover, we have

$$\mathbf{u}_n \to \mathbf{u} \text{ weakly in } L^2(0,T;W_0^{1,2}(\Omega;R^N)), \tag{7.56}$$

where the limit velocity $\mathbf{u}$ satisfies the no-slip boundary condition (7.7) in the sense of traces.

Since the convergence in (7.55) is strong, we also have

$$\varrho_n \mathbf{u}_n \to \varrho \mathbf{u} \quad \text{weakly-* in } L^\infty(0,T;L^{m_\infty}(\Omega;R^N)) \quad \text{with } m_\infty = \frac{2\gamma}{\gamma+1}. \quad (7.57)$$

Furthermore, by the same token,

$$\varrho_n Q(\vartheta_n) \to \varrho\overline{Q(\vartheta)} \quad \text{weakly in } L^2(0,T;L^q(\Omega)) \quad \text{with } 1 < q < \frac{2N\beta}{\beta(N-2)+2N}, \quad (7.58)$$

where

$$Q(\vartheta_n) \to \overline{Q(\vartheta)} \quad \text{weakly in } L^2(0,T;W^{1,2}(\Omega)). \quad (7.59)$$

Here we have used estimate (7.52) together with hypothesis (3.6).

In particular, we have proved that the limit functions ϱ, $\mathbf{u}$ solve problem (7.19) in $\mathcal{D}'((0,T) \times \Omega)$; more precisely, ϱ, $\mathbf{u}$ satisfy the integral identity (7.23). As a matter of fact, a much better result is available.

Lemma 7.5 *There exist $r > 1$ and $p > 2$ such that*

$$\partial_t \varrho_n, \Delta \varrho_n \text{ are bounded in } L^r((0,T) \times \Omega),$$

$$\nabla_x \varrho_n \text{ is bounded in } L^p(0,T;L^2(\Omega;R^N))$$

independently of n. Accordingly, the function ϱ belongs to the same class, satisfies equation (7.19) for a.a. $(t,\mathbf{x}) \in (0,T) \times \Omega$ together with the homogeneous Neumann boundary conditions in the sense of traces.

Proof We can write

$$\text{div}_x(\varrho_n \mathbf{u}_n) = \nabla_x \varrho_n \cdot \mathbf{u}_n + \varrho_n \, \text{div}_x \mathbf{u}_n$$

where, by virtue of estimates (7.49), (7.51),

$$\left\{ \begin{array}{c} \nabla_x \varrho_n \cdot \mathbf{u}_n \text{ is bounded in } L^1(0,T;L^{N/N-1}(\Omega)), \quad \text{for } N \geq 3, \\ \\ L^1(0,T;L^q(\Omega)) \quad \text{for any } q < 2 \text{ if } N = 2, \end{array} \right\}$$

and

$$\varrho_n \, \text{div}_x \mathbf{u}_n \text{ is bounded in } L^2(0,T;L^{2\beta/\beta+2}(\Omega;R^N)).$$

The idea is to apply the $L^p - L^q$ estimates stated in (7.21). To do this, however, we need an extra bit of information concerning integrability of the first term in time. Since

$$\varrho_n \mathbf{u}_n \text{ is bounded in } L^\infty(0, T; L^{2\beta/\beta+1}(\Omega; R^N))$$

$$\cap L^2(0, T; L^{2N\beta/(2N+\beta(N-2))}(\Omega; R^N))$$

for $N \geq 3$, and

$$\varrho_n \mathbf{u}_n \text{ is bounded in } L^\infty(0, T; L^{2\beta/\beta+1}(\Omega; R^N)) \cap L^2(0, T; L^q(\Omega; R^N))$$

for any $q < 2$ if $N = 2$, one can take

$$\beta > N$$

to obtain

$$\varrho_n \mathbf{u}_n \text{ is bounded in } L^p(0, T; L^2(\Omega; R^N)) \text{ for a certain } p > 2.$$

Indeed the interpolation inequality (for $N \geq 3$) yields

$$\|\varrho_n \mathbf{u}_n\|_{L^2(\Omega; R^N)} \leq \|\varrho_n \mathbf{u}_n\|_{L^{2\beta/\beta+1}(\Omega; R^N)}^{2(\beta-N)/(2\beta-N)} \|\varrho_n \mathbf{u}_n\|_{L^{2N\beta/2N+\beta(N-2)}(\Omega; R^N)}^{N/(2\beta-N)}$$

so we need $(2\beta - N)/N > 1$.

Now, the integral identity (7.23) may be interpreted as an abstract evolution equation of the form

$$\frac{\mathrm{d}}{\mathrm{d}t} v + \mathcal{A}[v] = g,$$

$$\mathcal{A} = -\varepsilon \Delta + \mathrm{Id} \text{ (+ the homogeneous Neumann boundary conditions)}$$

with $g \in L^p(0, T; \mathcal{D}(\mathcal{A}^{-1/2}))$, where $\mathcal{D}(\mathcal{A}^{1/2}) = W^{1,2}(\Omega)$.

Using the abstract semigroup theory of evolution equations, namely the maximal regularity estimates (see e.g. Chapter III.4 of [4]), we get

$$\varrho_n \text{ bounded in } L^p(0, T; \mathcal{D}(\mathcal{A}^{1/2})), \ \mathcal{D}(\mathcal{A}^{1/2}) = W^{1,2}(\Omega).$$

In particular, $\nabla_x \varrho_n$ belongs to the space $L^q(0, T; L^q(\Omega; R^N))$ for a certain $q > 2$.
Thus we have

$$\mathrm{div}_x(\varrho_n \mathbf{u}_n) \text{ bounded in } L^r((0, T) \times \Omega), \text{ with a certain } r > 1,$$

and the rest of the proof follows from the standard $(L^p - L^q)$ estimates (7.21).

$\square$

The estimates obtained in Lemma 7.5 together with those of Proposition 7.4 can be used to deduce from (7.19) that the integral mean functions

$$t \mapsto \int_\Omega \varrho_n \mathbf{u}_n \cdot \eta^j \, d\mathbf{x} \text{ form a precompact system in } C[0,T]$$

for any fixed j. This implies, by virtue of Corollary 2.1, that

$$\varrho_n \mathbf{u}_n \to \varrho\mathbf{u} \text{ in } C([0,T]; L^{2\gamma/\gamma+1}_{\text{weak}}(\Omega; R^N)). \tag{7.60}$$

In accordance with Theorem 2.8 and the hypothesis $\gamma > N/2$, the space $L^{2\gamma/\gamma+1}(\Omega)$ is compactly imbedded into $W^{-1,2}(\Omega)$, and, consequently,

$$\varrho_n \mathbf{u}_n \otimes \mathbf{u}_n \to \varrho\mathbf{u} \otimes \mathbf{u} \text{ weakly in } L^2(0,T; L^{c_2}(\Omega; R^{N^2})), \tag{7.61}$$

$$1 < c_2 \leq \frac{2N\gamma}{N + 2\gamma(N-2)}.$$

The functions ϱ_n and ϱ, being strong solutions of problem (7.19), they satisfy the energy equality

$$\|\varrho_n(t)\|^2_{L^2(\Omega)} + 2\varepsilon \int_0^t \|\nabla_x \varrho_n\|^2_{L^2(\Omega)} \, dt = -\int_0^t \int_\Omega \text{div}_x \mathbf{u}_n \varrho_n^2 \, d\mathbf{x} \, dt + \|\varrho_{0,\delta}\|^2_{L^2(\Omega)},$$

and

$$\|\varrho(t)\|^2_{L^2(\Omega)} + 2\varepsilon \int_0^t \|\nabla_x \varrho\|^2_{L^2(\Omega)} \, dt = -\int_0^t \int_\Omega \text{div}_x \mathbf{u}\varrho^2 \, d\mathbf{x} \, dt + \|\varrho_{0,\delta}\|^2_{L^2(\Omega)}$$

respectively. Consequently, we deduce

$$\|\nabla_x \varrho_n\|^2_{L^2((0,T)\times\Omega)} \quad \to \quad \|\nabla_x \varrho\|^2_{L^2((0,T)\times\Omega)}$$

and

$$\|\varrho_n(t)\|_{L^2(\Omega)} \to \|\varrho(t)\|_{L^2(\Omega)} \quad \text{for any } t \in [0,T]$$

which yields strong convergence

$$\nabla_x \varrho_n \to \nabla_x \varrho \text{ in } L^2((0,T) \times \Omega), \tag{7.62}$$

in particular,

$$\nabla_x \mathbf{u}_n \nabla_x \varrho_n \to \nabla_x \mathbf{u} \nabla_x \varrho \quad \text{in } [\mathcal{D}'((0,T) \times \Omega)]^N. \tag{7.63}$$

Finally, we can use Lemma 6.3 together with (7.58), (7.59), and equation (7.24) to obtain

$$(\delta + \varrho_n)Q(\vartheta_n) \to (\delta + \varrho)\overline{Q(\vartheta)} \text{ (strongly)} \quad \text{in } L^2(0,T; W^{-1,2}(\Omega))$$

provided $\beta > N - 1$. Consequently, in view of (7.59),

$$(\delta + \varrho_n)|Q(\vartheta_n)|^2 \to (\delta + \varrho)|\overline{Q(\vartheta)}|^2 \quad \text{in } \mathcal{D}'((0,T) \times \Omega). \tag{7.64}$$

As we have seen in Chapter 6, the function

$$\Phi(r,z) \mapsto \begin{cases} \frac{z^2}{r} & \text{if } r > 0, \\[2mm] 0 & \text{for } z = r = 0, \\[2mm] \infty & \text{if } r \le 0,\ z \ne 0 \end{cases}$$

is convex lower semi-continuous on R^2. Accordingly, by virtue of Corollary 2.2,

$$\int_0^T \int_\Omega \varrho |\overline{Q(\vartheta)}|^2 \, \mathrm{d}\mathbf{x}\, \mathrm{d}t \le \liminf_{n\to\infty} \int_0^T \int_\Omega (\varrho_n + \omega)|Q(\vartheta_n)|^2 \, \mathrm{d}\mathbf{x}\, \mathrm{d}t$$

for any $\omega > 0$, that is,

$$\int_0^T \int_\Omega \varrho |\overline{Q(\vartheta)}|^2 \, \mathrm{d}\mathbf{x}\, \mathrm{d}t \le \liminf_{n\to\infty} \int_0^T \int_\Omega \varrho_n |Q(\vartheta_n)|^2 \, \mathrm{d}\mathbf{x}\, \mathrm{d}t. \qquad (7.65)$$

Relation (7.65), together with (7.64), yields

$$\lim_{n\to\infty} \int_0^T \int_\Omega Q(\vartheta_n)^2 \, \mathrm{d}\mathbf{x}\, \mathrm{d}t = \int_0^T \int_\Omega \overline{Q(\vartheta)}^2 \, \mathrm{d}\mathbf{x}\, \mathrm{d}t;$$

whence we are allowed to conclude

$$Q(\vartheta_n) \to \overline{Q(\vartheta)} \ \text{(strongly)} \quad \text{in } L^2((0,T)\times\Omega).$$

As Q is strictly increasing, we get

$$\vartheta_n \to \vartheta \quad \text{in } L^2((0,T)\times\Omega). \qquad (7.66)$$

Indeed we have

$$0 = \lim_{n\to\infty} \int_0^T \int_\Omega (Q(\vartheta_n) - Q(\vartheta))(\vartheta_n - \vartheta) \, \mathrm{d}\mathbf{x}\, \mathrm{d}t$$

$$\ge \lim_{n\to\infty} c_v \int_0^T \int_\Omega |\vartheta_n - \vartheta|^2 \, \mathrm{d}\mathbf{x}\, \mathrm{d}t.$$

Now we are ready to state an existence result for problem (7.3)–(7.11).

Proposition 7.5 *Let $\Omega \subset R^N$ be a bounded domain of class $C^{2+\nu}$, $\nu > 0$. Let $\varepsilon > 0$, $\delta > 0$, and*

$$\beta > \max\{N, 4, \gamma\}$$

be fixed.

Then problem (7.3)–(7.11) admits at least one solution ϱ, **u**, ϑ in the following sense:

(1) *The density ϱ is a non-negative function such that*

$$\varrho \in L^r(0,T; W^{2,r}(\Omega)), \quad \partial_t \varrho \in L^r((0,T) \times \Omega) \quad \text{for a certain } r > 1,$$
(7.67)

the velocity **u** *belongs to the class* $L^2(0,T; W_0^{1,2}(\Omega; R^N))$, *equation (7.3) holds a.a. on* $(0,T) \times \Omega$, *and the boundary condition (7.4) as well as the initial condition (7.5) are satisfied in the sense of traces. Moreover, the total mass is conserved, specifically,*

$$\int_\Omega \varrho(t)\,dx = \int_\Omega \varrho_{0,\delta}\,dx \quad \text{for all } t \in [0,T];$$
(7.68)

and the following estimates hold:

$$\delta \int_0^T \int_\Omega \varrho^{\beta+1}\,dx\,dt \le c(\varepsilon),$$
(7.69)

$$\varepsilon \int_0^T \int_\Omega |\nabla_x \varrho|^2\,dx\,dt \le c \text{ with } c \text{ independent of } \varepsilon.$$
(7.70)

(2) *All quantities appearing in equation (7.6) are locally integrable, and the equation is satisfied in* $\mathcal{D}'((0,T) \times \Omega)$ *(in the sense of generalized derivatives). Moreover, we have*

$$\varrho\mathbf{u} \in C([0,T]; L_{weak}^{2\gamma/\gamma+1}(\Omega; R^N)),$$

and $\varrho\mathbf{u}$ satisfies the initial condition (7.8).

(3) *The energy inequality*

$$\int_\Omega \varrho(\tau)\Big(\frac{1}{2}|\mathbf{u}|^2 + P_m(\varrho) + \frac{\delta}{\beta-1}\varrho^{\beta-1}\Big)(\tau) + (\varrho+\delta)Q(\vartheta)(\tau)\,dx$$

$$+ \delta \int_0^\tau \int_\Omega \mathbb{S} : \nabla_x\mathbf{u} + \vartheta^{\alpha+1}\,dx\,dt$$

$$\le \int_\Omega \frac{|\mathbf{m}_0|^2}{\varrho_{0,\delta}} + \varrho_{0,\delta}P_m(\varrho_{0,\delta}) + \frac{\delta}{\beta-1}\varrho_{0,\delta}^\beta + (\varrho_{0,\delta}+\delta)Q(\vartheta_{0,\delta})\,dx$$

$$+ \int_0^\tau \int_\Omega p_b(\varrho)\,\text{div}_x\mathbf{u} + \varrho\mathbf{f}\cdot\mathbf{u}\,dx\,dt$$
(7.71)

holds for a.a. $\tau \in [0,T]$.

(4) *The temperature ϑ is a non-negative function,*

$$\vartheta \in L^{\alpha+1}((0,T) \times \Omega), \ \ \vartheta^{\alpha/2} \in L^2(0,T;W^{1,2}(\Omega,R^N)),$$

satisfying an integral inequality

$$\int_0^T \int_\Omega \left((\delta + \varrho)Q_{\mathrm{h}}(\vartheta) \right) \partial_t \varphi + \varrho Q_{\mathrm{h}}(\vartheta)\mathbf{u} \cdot \nabla_x \varphi + \mathcal{K}_{\mathrm{h}}(\vartheta)\Delta\varphi$$

$$- \delta\vartheta^{\alpha+1}h(\vartheta)\varphi \, \mathrm{d}\mathbf{x} \, \mathrm{d}t$$

$$\leq \int_0^T \int_\Omega \left((\delta - 1)h(\vartheta)\mathbb{S} : \nabla_x \mathbf{u} + \kappa(\vartheta)h'(\vartheta)|\nabla_x \vartheta|^2 \right)\varphi \, \mathrm{d}\mathbf{x} \, \mathrm{d}t$$

$$+ \int_0^T \int_\Omega \vartheta h(\vartheta)p_\vartheta(\varrho) \, \mathrm{div}_x \mathbf{u} \ \varphi \, \mathrm{d}\mathbf{x} \, \mathrm{d}t$$

$$+ \varepsilon \int_0^T \int_\Omega \nabla_x \varrho \ \cdot \nabla_x \left[\left(Q_{\mathrm{h}}(\vartheta) - Q(\vartheta)h(\vartheta) \right)\varphi \right] \mathrm{d}\mathbf{x} \, \mathrm{d}t$$

$$- \int_\Omega (\delta + \varrho_{0,\delta})Q_{\mathrm{h}}(\vartheta_{0,\delta})\varphi(0) \, \mathrm{d}\mathbf{x} \qquad (7.72)$$

for any h as in (4.48),and any test function φ,

$$\varphi \geq 0, \quad \varphi \in C^2([0,T] \times \overline{\Omega}), \quad \nabla_x \varphi \cdot \mathbf{n}|_{\partial\Omega} = 0, \quad \varphi(T) = 0. \qquad (7.73)$$

Here $\mathcal{K}_{\mathrm{h}}$, Q_{h} are determined through (4.49).

Remark   In view of future applications, it seems more convenient to write the energy inequality in a "weak" form:

$$\int_0^T \int_\Omega (-\partial_t \psi) \left[\varrho\left(\frac{1}{2}|\mathbf{u}|^2 + P_{\mathrm{m}}(\varrho) + \frac{\delta}{\beta - 1}\varrho^{\beta-1}\right) + (\varrho + \delta)Q(\vartheta)(\tau) \right] \mathrm{d}\mathbf{x} \, \mathrm{d}t$$

$$+ \delta \int_0^T \int_\Omega \psi(\mathbb{S} : \nabla_x \mathbf{u} + \vartheta^{\alpha+1}) \, \mathrm{d}\mathbf{x} \, \mathrm{d}t$$

$$\leq \int_\Omega \frac{|\mathbf{m}_0|^2}{\varrho_{0,\delta}} + \varrho_{0,\delta}P_{\mathrm{m}}(\varrho_{0,\delta}) + \frac{\delta}{\beta - 1}\varrho_{0,\delta}^\beta + (\varrho_{0,\delta} + \delta)Q(\vartheta_{0,\delta}) \, \mathrm{d}\mathbf{x}$$

$$+ \int_0^T \int_\Omega \psi\Big(p_b(\varrho) \, \mathrm{div}_x \mathbf{u} + \varrho\mathbf{f} \cdot \mathbf{u}\Big) \mathrm{d}\mathbf{x} \, \mathrm{d}t \qquad (7.74)$$

for any test function

$$\psi \in C^\infty[0,T], \quad \psi(0) = 1, \quad \partial_t \psi \leq 0.$$

Remark   Relation (7.72) is nothing other than a "renormalized and regularized" thermal energy inequality introduced in Definition 4.4.

Proof (i) Clearly, the part concerning problem (7.3)–(7.5) follows directly from Lemma 7.5, the relations (7.55), (7.57) together with (7.42) and the estimates (7.49) stated in Proposition 7.4.

(ii) Similarly, fixing $\eta = \eta_j$ in (7.38), one can multiply this equation by a test function $\psi \in \mathcal{D}(0, T)$, integrate by parts with respect to t, and pass to the limit for $n \to \infty$, to obtain that (7.6) holds in $\mathcal{D}'((0, T) \times \Omega)$. Here, of course, we have used (7.60), (7.61) together with (7.55), (7.56), (7.66), and (7.63).

(iii) As for the energy inequality (7.71), we claim it follows from (7.43). Indeed we have

$$\int_0^\tau \int_\Omega \mathbb{S} : \nabla_x \mathbf{u} + \vartheta^{\alpha+1} \, d\mathbf{x} \, dt \leq \liminf_{n \to \infty} \int_0^\tau \int_\Omega \mathbb{S}_n : \nabla_x \mathbf{u}_n + \vartheta_n^{\alpha+1} \, d\mathbf{x} \, dt,$$

and

$$\int_0^\tau \int_\Omega p_{\mathrm{b}}(\varrho) \, \mathrm{div}_x \mathbf{u} + \varrho \mathbf{f} \cdot \mathbf{u} \, d\mathbf{x} \, dt = \lim_{n \to \infty} \int_0^\tau \int_\Omega p_{\mathrm{b}}(\varrho_n) \, \mathrm{div}_x \mathbf{u}_n + \varrho_n \mathbf{f} \cdot \mathbf{u}_n \, d\mathbf{x} \, dt$$

for any $\tau \geq 0$ in accordance with (7.55), (7.56).

On the other hand, one can use (7.44), (7.55), (7.58), and (7.61) to deduce

$$\int_\Omega \varrho |\mathbf{u}|^2 (\tau+) \eta \, d\mathbf{x} = \lim_{d \to 0+} \frac{1}{d} \int_\tau^{\tau+d} \int_\Omega \varrho |\mathbf{u}|^2 \eta \, d\mathbf{x} \, dt$$

$$= \lim_{d \to 0+} \frac{1}{d} \left(\lim_{n \to \infty} \int_\tau^{\tau+d} \int_\Omega \varrho_n |\mathbf{u}_n|^2 \eta \, d\mathbf{x} \, dt \right) \quad \text{for any } \eta \in \mathcal{D}(\Omega),$$

and, similarly,

$$\int_\Omega P_{\mathrm{m}}(\varrho)(\tau+) \eta \, d\mathbf{x} = \lim_{d \to 0+} \frac{1}{d} \left(\lim_{n \to \infty} \int_\tau^{\tau+d} \int_\Omega P_{\mathrm{m}}(\varrho_n) \eta \, d\mathbf{x} \, dt \right),$$

$$\int_\Omega \varrho^\beta(\tau+) \eta \, d\mathbf{x} = \lim_{d \to 0+} \frac{1}{d} \left(\lim_{n \to \infty} \int_\tau^{\tau+d} \int_\Omega \varrho_n^\beta \eta \, d\mathbf{x} \, dt \right),$$

with

$$\int_\Omega (\delta \vartheta + \varrho Q(\vartheta))(\tau+) \, d\mathbf{x} = \lim_{d \to 0+} \frac{1}{d} \left(\lim_{n \to \infty} \int_\tau^{\tau+d} \int_\Omega (\delta \vartheta_n + \varrho_n Q(\vartheta_n)) \, d\mathbf{x} \, dt \right)$$

for a.e. $\tau \in [0, T]$.

These relations, together with (7.43) yield the desired inequality (7.71).

(iv) Our final task is to pass to the limit for $n \to \infty$ in (7.47) to obtain (7.72). Note that it is enough to show that one can pass to the limit in all non-linear terms contained in (7.47).

To begin with, observe that, since h satisfies (4.48),

$$\lim_{z \to \infty} h(z) = 0.$$

Thus

$$\lim_{z \to \infty} \frac{Q_{\mathrm{h}}(z)}{Q(z)} = 0,$$

and, consequently, we can use (7.55), (7.66) together with estimates (7.52) to deduce

$$(\delta + \varrho_n)Q_{\mathrm{h}}(\vartheta_n) \to (\delta + \varrho)Q_{\mathrm{h}}(\vartheta) \quad \text{in } L^1((0,T) \times \Omega),$$

$$\varrho_n Q_{\mathrm{h}}(\vartheta_n)\mathbf{u}_n \to \varrho Q_{\mathrm{h}}(\vartheta)\mathbf{u} \text{ weakly} \quad \text{in } L^r((0,T) \times \Omega),$$

and,

$$\vartheta_n h(\vartheta_n)\, p_\vartheta(\varrho_n)\, \mathrm{div}_x \mathbf{u}_n \to \vartheta h(\vartheta)\, p_\vartheta(\varrho)\, \mathrm{div}_x \mathbf{u} \text{ weakly} \quad \text{in } L^r((0,T) \times \Omega)$$

for a certain $r > 1$.

Moreover, because of convexity of the function

$$[\mathbb{M}, \vartheta] \mapsto \begin{cases} h(\vartheta)\left(\frac{\mu}{2}\mathbb{M} : \mathbb{M} + \lambda\,(\mathrm{tr}[\mathbb{M}])^2\right) & \text{if } \vartheta \geq 0,\ \mathbb{M} \in R^{N^2}, \\ \infty & \text{if } \vartheta < 0 \end{cases}$$

(cf. Lemma 4.8), we get

$$\int_0^T \int_\Omega h(\vartheta)\,\mathbb{S} : \nabla_x \mathbf{u}\ \varphi \,\mathrm{dx}\,\mathrm{dt} \leq \liminf_{n \to \infty} \int_0^T \int_\Omega h(\vartheta_n)\,\mathbb{S}_n : \nabla_x \mathbf{u}_n\ \varphi \,\mathrm{dx}\,\mathrm{dt}$$

for any non-negative test function φ. Similarly,

$$-\int_0^T \int_\Omega \kappa(\vartheta)h'(\vartheta)|\nabla_x \vartheta|^2 \varphi \,\mathrm{dx}\,\mathrm{dt} \leq -\liminf_{n \to \infty} \int_0^T \int_\Omega \kappa(\vartheta_n)h'(\vartheta_n)|\nabla_x \vartheta_n|^2 \varphi \,\mathrm{dx}\,\mathrm{dt}$$

(remember $h' \leq 0$).

Now, because of strong convergence of $\nabla_x \varrho_n$ established in (7.62), we have

$$\int_0^T \int_\Omega \nabla_x \varrho_n \cdot \nabla_x \Big[\Big(Q_{\mathrm{h}}(\vartheta_n) - Q(\vartheta_n)h(\vartheta_n)\Big)\varphi\Big] \,\mathrm{dx}\,\mathrm{dt}$$

$$\to \int_0^T \int_\Omega \nabla_x \varrho \cdot \nabla_x \Big[\Big(Q_{\mathrm{h}}(\vartheta) - Q(\vartheta)h(\vartheta)\Big)\varphi\Big] \,\mathrm{dx}\,\mathrm{dt} \quad \text{for } n \to \infty.$$

Finally, by virtue of (7.52), (7.66),

$$\mathcal{K}_{\mathrm{h}}(\vartheta_n) \to \mathcal{K}_{\mathrm{h}}(\vartheta) \quad \text{in } L^1((0,T) \times \Omega),$$

$$h(\vartheta_n)\vartheta_n^{\alpha+1} \to h(\vartheta)\vartheta^{\alpha+1} \quad \text{in } L^1((0,T) \times \Omega),$$

which, together with above relations, completes the proof of inequality (7.72).

$\square$

7.4 Vanishing artificial viscosity

Our next goal is to let $\varepsilon \to 0$ in the modified continuity equation (7.3). To this end, let us denote by ϱ_ε, $\mathbf{u}_\varepsilon$, ϑ_ε the corresponding solution of the *approximate problems* the existence of which was stated in Proposition 7.5. At this stage of the proof of Theorem 7.1, we definitely loose boundedness of $\nabla_x \varrho_\varepsilon$ and, consequently, strong compactness of the sequence $\{\varrho_\varepsilon\}_{\varepsilon>0}$ in $L^1((0,T) \times \Omega)$ becomes a central issue. On the other hand, with $\delta > 0$ fixed, the modified thermal energy equation (7.9) will be treated in a similar way as in Section 7.3.

7.4.1 Local pressure estimates

We evoke the *local pressure estimates* discussed in Chapter 5. Since the data $\varrho_{0,\delta}$, $\mathbf{m}_{0,\delta}$, and $\vartheta_{0,\delta}$ are fixed, the energy inequality (7.71) renders

$$\varrho_\varepsilon \text{ bounded } \text{ in } L^\infty(0,T; L^\beta(\Omega)),$$

and

$$\mathbf{u}_\varepsilon \text{ bounded } \text{ in } L^2(0,T; W_0^{1,2}(\Omega; R^N)),$$

which, together with Theorem 2.6 and Hölder's inequality (2.1), yields

$$\varrho_\varepsilon \mathbf{u}_\varepsilon \text{ bounded in } L^2(0,T; L^q(\Omega; R^N)), \text{ with } q > 2, \qquad (7.75)$$

provided $\beta > N$.

Lemma 7.6 *For any compact $O \subset ((0,T) \times \Omega)$, there is a constant $c = c(\delta, O)$ independent of ε such that*

$$\delta \int_O \varrho_\varepsilon^{\beta+1} \, \mathrm{dx} \, \mathrm{dt} \le c(\delta, O). \qquad (7.76)$$

Proof Although the heuristic principle is the same as in Chapter 5, the proof of (7.76) requires a slight modification to accommodate the extra terms in (7.3) and (7.6).

Set

$$B_\omega = [\varrho_\varepsilon]_x^\omega,$$

where $v \mapsto [v]_x^\omega$ are the smoothing operators introduced in (2.4). In accordance with Lemma 7.5, the functions ϱ_ε, $\mathbf{u}_\varepsilon$ satisfy (7.3) a.a. on $(0,T) \times \Omega$ together with the boundary conditions (7.4), in particular,

$$\partial_t \varrho_\varepsilon + \mathrm{div}_x(\varrho_\varepsilon \mathbf{u}_\varepsilon) = \varepsilon \, \mathrm{div}_x(1_\Omega \nabla_x \varrho_\varepsilon) \text{ in } \mathcal{D}'((0,T) \times R^N) \qquad (7.77)$$

provided ϱ_ε, $\mathbf{u}_\varepsilon$ were extended to be zero outside Ω. Here, the symbol $1_\Omega \nabla_x \varrho_\varepsilon$ denotes the function

$$
1_\Omega(\mathbf{x}) \nabla_x \varrho_\varepsilon =
\begin{cases}
\nabla_x \varrho_\varepsilon(\mathbf{x}) & \text{if } \mathbf{x} \in \Omega, \\[2mm]
0 & \text{if } \mathbf{x} \in R^N \setminus \Omega.
\end{cases}
$$

Consequently, we have

$$
\partial_t B_\omega + \operatorname{div}_x(B_\omega \mathbf{u}_\varepsilon) = h_\omega \quad \text{in } \mathcal{D}'((0,T) \times R^N)
$$

with

$$
h_\omega = \operatorname{div}_x(B_\omega \mathbf{u}_\varepsilon) - \operatorname{div}_x([\varrho_\varepsilon \mathbf{u}_\varepsilon]_x^\omega) + \varepsilon \operatorname{div}_x[1_\Omega \nabla_x \varrho_\varepsilon]_x^\omega.
$$

By virtue of (7.70), (7.75), the functions ϱ_ε, $\mathbf{u}_\varepsilon$, B_ω, h_ω and

$$
p_\varepsilon = p_e(\varrho_\varepsilon) + \vartheta_\varepsilon p_\vartheta(\varrho_\varepsilon)
$$

satisfy the hypotheses of Lemma 5.3 with $q = 2$. Note that, by virtue of the energy inequality (7.71), the temperature

$$
\vartheta_\varepsilon \text{ is bounded in the space } L^{\alpha+1}((0,T) \times \Omega)
$$

while, in accordance with hypothesis (3.2),

$$
p_\vartheta(\varrho_\varepsilon) \text{ is bounded} \quad \text{in } L^\infty(0,T; L^N(\Omega)),
$$

in particular, the thermal pressure

$$
\vartheta_\varepsilon p_\vartheta(\varrho_\varepsilon) \text{ is bounded} \quad \text{in } L^1((0,T) \times \Omega)
$$

by a constant independent of ε.

As a matter of fact, equation (7.6) contains the extra term $\varepsilon \nabla_x \mathbf{u}_\varepsilon \nabla_x \varrho_\varepsilon$, which is, however, bounded in $L^1((0,T) \times \Omega)$ and does not cause any additional problem in the proof of Lemma 5.3.

Thus we are allowed to use the conclusion of Lemma 5.3 to obtain

$$
\int_0^T \int_\Omega \psi \eta \Big(\xi p_\varepsilon B_\omega - \mathbb{S}_\varepsilon : (\nabla_x \Delta^{-1} \nabla_x)[\xi B_\omega] \Big) \, \mathrm{d}\mathbf{x} \, \mathrm{d}t = \sum_{j=1}^{10} I_j, \tag{7.78}
$$

where

$$I_1 = \int_0^T \int_\Omega \psi (\mathbb{S}_\varepsilon \nabla_x \eta) \cdot \mathcal{A}[\xi B_\omega] \, dx \, dt,$$

$$I_2 = -\int_0^T \int_\Omega \psi \, p_\varepsilon \, \nabla_x \eta \cdot \mathcal{A}[\xi B_\omega] \, dx \, dt,$$

$$I_3 = -\int_0^T \int_\Omega \psi \eta \, \varrho_\varepsilon \mathbf{f} \cdot \mathcal{A}[\xi B_\omega] \, dx \, dt,$$

$$I_4 = -\int_0^T \int_\Omega \psi ([\varrho_\varepsilon \mathbf{u}_\varepsilon \otimes \mathbf{u}_\varepsilon] \nabla_x \eta) \cdot \mathcal{A}[\xi B_\omega] \, dx \, dt,$$

$$I_5 = -\int_0^T \int_\Omega \psi \eta \, \varrho_\varepsilon \mathbf{u}_\varepsilon \cdot \mathcal{A}[\xi B_\omega] \, dx \, dt,$$

$$I_6 = -\int_0^T \int_\Omega \partial_t \psi \, \eta \, \varrho_\varepsilon \mathbf{u}_\varepsilon \cdot \mathcal{A}[\xi B_\omega] \, dx \, dt,$$

$$I_7 = -\int_0^T \int_\Omega \psi \eta \, \varrho_\varepsilon \mathbf{u}_\varepsilon \cdot \mathcal{A}[\xi h_\omega] \, dx \, dt,$$

$$I_8 = -\int_0^T \int_\Omega \psi \mathbf{u}_\varepsilon \cdot (\nabla_x \Delta^{-1} \nabla_x)[\xi B_\omega] \eta \varrho_\varepsilon \mathbf{u}_\varepsilon \, dx \, dt,$$

$$I_9 = \int_0^T \int_\Omega \psi \xi \, B_\omega \mathbf{u}_\varepsilon \cdot (\nabla_x \Delta^{-1} \mathrm{div}_x)[\eta \varrho_\varepsilon \mathbf{u}_\varepsilon] \, dx \, dt,$$

and, in addition,

$$I_{10} = \varepsilon \int_0^T \int_\Omega \psi \eta \, \nabla_x \mathbf{u}_\varepsilon \nabla_x \varrho_\varepsilon \cdot \mathcal{A}[\xi B_\omega] \, dx \, dt,$$

where $\mathcal{A} = \nabla_x \Delta^{-1}$ is the integral operator introduced in (5.4), and $\psi \in \mathcal{D}(0, T)$, η, $\xi \in \mathcal{D}(\Omega)$ are arbitrary test functions such that

$$\psi(t)\eta(\mathbf{x})\xi(\mathbf{x}) = 1 \quad \text{for all } (t, \mathbf{x}) \in O.$$

Now, by virtue of Lemma 5.2 and Theorem 2.6,

$$\mathcal{A}[\xi B_\omega] \text{ are bounded in } [L^\infty((0, T) \times \Omega)]^N$$

provided $\beta > N$, where the bound is independent of the parameters ω and ε. Consequently, the integrals I_1–I_6 are bounded by a constant independent of ω and ε.

Next, by the same token,

$$\xi h_\omega \text{ are bounded in } L^2(0,T;W^{-1,2}(\Omega)),$$

$$\mathcal{A}[\xi h_\omega] \text{ bounded in } L^2((0,T) \times \Omega)$$

again for $\beta > N$. This relation together with (7.75) yields a bound on I_7 independent of ε, ω. Moreover, similar arguments can be used to show boundedness of I_8 and I_9.

Furthermore, we have

$$|I_{10}| \le \varepsilon \|\psi\eta\|_{L^\infty((0,T)\times\Omega)} \|\mathbf{u}_\varepsilon\|_{L^2((0,T)\times\Omega)} \|\varrho_\varepsilon\|_{L^2((0,T)\times\Omega)} \|\mathcal{A}[\xi B_\omega]\|_{L^\infty((0,T)\times\Omega)},$$

with the right-hand side bounded in accordance with estimates (7.70), (7.71).

Finally, observe that the integral

$$\int_0^T \int_\Omega \mathbb{S}_\varepsilon : (\nabla_x \Delta^{-1} \nabla_x)[\xi B_\omega] \, d\mathbf{x} \, dt$$

is bounded provided $\beta \ge 2$.

In accordance with (7.69), we are allowed to let $\omega \to 0$ to obtain (7.76). $\square$

7.4.2 Vanishing viscosity limit

In accordance with the estimates established in Proposition 7.5, specifically the energy inequality (7.71), we can assume that

$$\varrho_\varepsilon \to \varrho \text{ weakly(-*) in } L^\infty(0,T;L^\beta(\Omega)),$$

$$\mathbf{u}_\varepsilon \to \mathbf{u} \text{ weakly in } L^2(0,T;W_0^{1,2}(\Omega;R^N)), \tag{7.79}$$

and

$$\vartheta_\varepsilon \to \vartheta \text{ weakly in } L^{\alpha+1}((0,T) \times \Omega).$$

Furthermore, we can take

$$h(\vartheta) = \frac{1}{(1+\vartheta)^\omega}, \quad \omega \in (0,1),$$

in the integral inequality (7.72) to obtain

$$\int_0^T \int_\Omega \psi \left| \nabla_x \vartheta_\varepsilon^{(\alpha+1-\omega)/2} \right|^2 \, d\mathbf{x} \, dt \le c(\omega) \sum_{j=1}^4 I_j, \tag{7.80}$$

where

$$I_1 = \delta \int_0^T \int_\Omega \psi \, \vartheta_\varepsilon^{\alpha+1} \, \mathrm{dx} \, \mathrm{dt},$$

$$I_2 = -\int_0^T \int_\Omega \partial_t \psi \left((\delta + \varrho_\varepsilon) Q_\mathrm{h}(\vartheta_\varepsilon) \right) \mathrm{dx} \, \mathrm{dt},$$

$$I_3 = \int_0^T \int_\Omega \psi \, \vartheta_\varepsilon p_\vartheta(\varrho_\varepsilon) \, |\mathrm{div}_x \mathbf{u}_\varepsilon| \, \mathrm{dx} \, \mathrm{dt},$$

and

$$I_4 = \varepsilon \int_0^T \int_\Omega \psi \, \nabla_x \varrho_\varepsilon \nabla_x \left(Q_\mathrm{h}(\vartheta_\varepsilon) - Q(\vartheta_\varepsilon) \frac{1}{(1 + \vartheta_\varepsilon)^\omega} \right) \mathrm{dx} \, \mathrm{dt},$$

with

$$\psi \in \mathcal{D}(0, T), \quad 0 \le \psi \le 1.$$

It is easy to see that I_1 and I_2 are bounded by a constant independent of ε in view of the energy estimates (7.71) provided ψ has at most one local maximum on $[0, T]$. Indeed we have

$$|I_2| \le 2\mathrm{ess} \sup_{t \in [0,T]} \left\{ \int_\Omega (\delta + \varrho_\varepsilon) Q_\mathrm{h}(\vartheta_\varepsilon)(t) \right\} \max_{t \in [0,T]} \psi(t).$$

Moreover, by virtue of Theorem 2.6 and Hölder's inequality (2.1),

$$|I_3| \le c \|\vartheta_\varepsilon\|_{L^2(0,T;W^{1,2}(\Omega))} \|\mathbf{u}_\varepsilon\|_{L^2(0,T;W_0^{1,2}(\Omega;R^N))} \|p_\vartheta(\varrho_\varepsilon)\|_{L^\infty(0,T;L^q(\Omega))},$$

where

$$q > 2 \quad \text{if } N = 2, \quad q = N \text{ for } N \ge 3,$$

and

$$|I_4| \le \varepsilon \left\| \nabla_x \varrho_\varepsilon \right\|_{L^2((0,T) \times \Omega)} \left\| \frac{Q(\vartheta_\varepsilon)}{(1 + \vartheta_\varepsilon)^{\omega+1}} \nabla_x \vartheta_\varepsilon \right\|_{L^2((0,T) \times \Omega)}.$$

As the functions Q and p_ϑ comply with the structural conditions specified in hypotheses (3.2), (3.5), and (3.6), we can use the *energy estimates* (7.71) to conclude that the left-hand side of the inequality (7.80) is bounded by a constant independent of ε:

$$\|\vartheta_\varepsilon^{(\alpha+1-\omega)/2}\|_{L^2(0,T;W^{1,2}(\Omega))} \le c(\omega) \quad \text{for any } \omega > 0. \tag{7.81}$$

In particular, we can assume that

$$Q(\vartheta_\varepsilon) \to \overline{Q(\vartheta)} \quad \text{weakly in } L^2(0, T; W^{1,2}(\Omega)).$$

Note that we already have ϑ_ε bounded in $L^{\alpha+1}((0,T) \times \Omega)$.

Next, we have

$$\varepsilon \operatorname{div}_x(1_\Omega \nabla_x \varrho_\varepsilon) \to 0 \text{ in } L^2(0,T;W^{-1,2}(R^N)); \tag{7.82}$$

therefore

$$\varrho_\varepsilon \to \varrho \text{ in } C([0,T]; L^\beta_{\text{weak}}(\Omega)) \tag{7.83}$$

(see Corollary 2.1).

Moreover,

$$\varepsilon \nabla_x \mathbf{u}_\varepsilon \nabla_x \varrho_\varepsilon \to 0 \text{ in } L^1(0,T; L^1(\Omega; R^N)). \tag{7.84}$$

As ϱ_ε, $\mathbf{u}_\varepsilon$ satisfy the energy inequality (7.71), the momenta $(\varrho\mathbf{u})_\varepsilon$ are bounded in $L^\infty(0,T; L^{\frac{2\beta}{\beta+1}}(\Omega; R^N))$; whence

$$(\varrho\mathbf{u})_\varepsilon \to \varrho\mathbf{u} \text{ in } C([0,T]; L^{\frac{2\beta}{\beta+1}}_{\text{weak}}(R^N; R^N))$$

provided ϱ_ε, $\mathbf{u}_\varepsilon$ were extended to be zero outside Ω. Indeed note that the extra terms $\varepsilon \nabla_x \mathbf{u}_\varepsilon \nabla_x \varrho$ multiplied by an arbitrary test function $\eta \in C(\overline{\Omega})$ and integrated over Ω tend to zero for $\varepsilon \to 0$ uniformly in $t \in [0,T]$.

In particular, we have shown that the limit functions ϱ, $\mathbf{u}$ satisfy the continuity equation

$$\partial_t \varrho + \operatorname{div}_x(\varrho\mathbf{u}) = 0 \text{ in } \mathcal{D}'((0,T) \times R^N). \tag{7.85}$$

Moreover, as the Lebesgue space $L^{\frac{2\beta}{\beta+1}}(\Omega)$ is compactly imbedded into $W^{-1,2}(\Omega)$ as soon as $\beta > N/2$, we infer that

$$\varrho_n \mathbf{u}_n \otimes \mathbf{u}_n \to \varrho\mathbf{u} \otimes \mathbf{u} \text{ weakly in } L^2(0,T; L^{c_2}(\Omega; R^{N^2})),$$

where the exponent $c_2 > 1$ can be taken as in Theorem 3.1.

Denoting by

$$\overline{p}, \text{ a weak limit of } p_e(\varrho_\varepsilon) + \delta\varrho_\varepsilon^\beta + \vartheta_\varepsilon p_\vartheta(\varrho_\varepsilon) \text{ in } L^{\beta+1/\beta}((0,T) \times \Omega), \tag{7.86}$$

we recover the momentum equation in the form

$$\partial_t(\varrho\mathbf{u}) + \operatorname{div}_x(\varrho\mathbf{u} \otimes \mathbf{u}) + \nabla_x \overline{p} = \operatorname{div}_x \mathbb{S} + \varrho\mathbf{f} \text{ in } \mathcal{D}'((0,T) \times \Omega). \tag{7.87}$$

Note that we can always take $\beta \geq \gamma$, and that the thermal pressure

$$\vartheta_\varepsilon p_\vartheta(\varrho_\varepsilon) \text{ is bounded in } L^2((0,T) \times \Omega)$$

in accordance with energy estimates (7.71), (7.81), and hypothesis (3.2).

7.4.3 Strong convergence of densities

The next step will be to show strong convergence of the sequence of densities $\{\varrho_\varepsilon\}_{\varepsilon>0}$ in $L^1((0,T) \times \Omega)$. Unlike in Section 7.3, where estimates on $\nabla_x \varrho$ were available, this is a difficult task. To this end, we adapt the proof of Proposition 6.1 together with the estimates on density oscillations discussed in Section 6.6.

First of all, one can pass to the limit for $\omega \to 0$ in (7.78) to obtain

$$\int_0^T \int_\Omega \psi\eta\Big(\xi p_\varepsilon \varrho_\varepsilon - \mathbb{S}_\varepsilon : (\nabla_x \Delta^{-1}\nabla_x)[\xi\varrho_\varepsilon]\Big)\,\mathrm{d}\mathbf{x}\,\mathrm{d}t = \sum_{j=1}^8 I_j$$

$$+ \int_0^T \int_\Omega \psi\mathbf{u}_\varepsilon \Big[\xi\varrho_\varepsilon(\nabla_x \Delta^{-1}\nabla_x)[\eta\varrho_\varepsilon \mathbf{u}_\varepsilon] - (\nabla_x \Delta^{-1}\nabla_x)[\xi\varrho_\varepsilon]\eta\varrho_\varepsilon \mathbf{u}_\varepsilon\Big]\,\mathrm{d}\mathbf{x}\,\mathrm{d}t$$

$$(7.88)$$

where

$$I_1 = \int_0^T \int_\Omega \psi(\mathbb{S}_\varepsilon \nabla_x \eta) \cdot \mathcal{A}[\xi\varrho_\varepsilon]\,\mathrm{d}\mathbf{x}\,\mathrm{d}t,$$

$$I_2 = -\int_0^T \int_\Omega \psi\, p_\varepsilon \nabla_x \eta \cdot \mathcal{A}[\xi\varrho_\varepsilon]\,\mathrm{d}\mathbf{x}\,\mathrm{d}t,$$

$$I_3 = -\int_0^T \int_\Omega \psi\eta\, \varrho_\varepsilon \mathbf{f} \cdot \mathcal{A}[\xi\varrho_\varepsilon]\,\mathrm{d}\mathbf{x}\,\mathrm{d}t,$$

$$I_4 = -\int_0^T \int_\Omega \psi([\varrho_\varepsilon \mathbf{u}_\varepsilon \otimes \mathbf{u}_\varepsilon]\nabla_x \eta) \cdot \mathcal{A}[\xi\varrho_\varepsilon]\,\mathrm{d}\mathbf{x}\,\mathrm{d}t,$$

$$I_5 = -\int_0^T \int_\Omega \psi\eta\, \varrho_\varepsilon \mathbf{u}_\varepsilon \cdot \mathcal{A}[\xi\varrho_\varepsilon]\,\mathrm{d}\mathbf{x}\,\mathrm{d}t,$$

$$I_6 = -\int_0^T \int_\Omega \partial_t\psi\, \eta\, \varrho_\varepsilon \mathbf{u}_\varepsilon \cdot \mathcal{A}[\xi\varrho_\varepsilon]\,\mathrm{d}\mathbf{x}\,\mathrm{d}t,$$

$$I_7 = -\varepsilon \int_0^T \int_\Omega \psi\eta\, \varrho_\varepsilon \mathbf{u}_\varepsilon \cdot \mathcal{A}[\xi\mathrm{div}_x(1_\Omega \nabla_x \varrho_\varepsilon)]\,\mathrm{d}\mathbf{x}\,\mathrm{d}t,$$

and

$$I_8 = \varepsilon \int_0^T \int_\Omega \psi\eta\, \nabla_x \mathbf{u}_\varepsilon \nabla_x \varrho_\varepsilon \cdot \mathcal{A}[\xi\varrho_\varepsilon]\,\mathrm{d}\mathbf{x}\,\mathrm{d}t.$$

Formula (7.88) is formally the same as (6.8), with $B(\varrho_n)$ replaced by ϱ_ε, and except for the integrals I_7, I_8, for which we have

$$|I_7| + |I_8| \to 0 \quad \text{for } \varepsilon \to 0$$

in accordance with (7.70),(7.75), (7.82), and (7.84).

Consequently, starting from formula (6.8) one can repeat step by step the proof of Proposition 6.1 to obtain the following result on the weak continuity of the effective viscous flux.

Lemma 7.7 *We have*

$$\lim_{\varepsilon \to 0} \int_0^T \int_\Omega \psi \eta \Big(p_e(\varrho_\varepsilon) + \delta \varrho^\beta + \vartheta_\varepsilon p_\vartheta(\varrho_\varepsilon) - (\lambda + 2\mu) \mathrm{div}_x \mathbf{u}_\varepsilon \Big) \varrho_\varepsilon \, \mathrm{d}x \, \mathrm{d}t$$

$$= \int_0^T \int_\Omega \psi \eta \Big(\overline{p} - (\lambda + 2\mu) \mathrm{div}_x \mathbf{u} \Big) \varrho \, \mathrm{d}x \, \mathrm{d}t$$

for any $\psi \in \mathcal{D}(0,T)$, $\eta \in \mathcal{D}(\Omega)$ *provided*

$$\beta > \max\{N, 4, \gamma\}.$$

Now, since the limit functions satisfy the continuity equation (7.85) in $\mathcal{D}'((0,T) \times R^N)$ and ϱ belongs to $L^2((0,T) \times R^N)$, we can use Corollary 4.1 to deduce that ϱ is a renormalized solution of (7.85) in the sense of Definition 4.1. In particular,

$$\partial_t(\varrho \log(\varrho)) + \mathrm{div}_x(\varrho \log(\varrho)\mathbf{u}) + \varrho \, \mathrm{div}_x \mathbf{u} = 0 \text{ in } \mathcal{D}'((0,T) \times R^N). \qquad (7.89)$$

On the other hand, by virtue of Lemma 7.5, ϱ_ε satisfies (7.3) a.a. on the set $(0,T) \times \Omega$. Thus we are allowed to multiply (7.3) on $B'(\varrho_\varepsilon)$ to obtain

$$\partial_t B(\varrho_\varepsilon) + \mathrm{div}_x(B(\varrho_\varepsilon)\mathbf{u}_\varepsilon) + \Big(B'(\varrho_\varepsilon)\varrho_\varepsilon - B(\varrho_\varepsilon) \Big) \mathrm{div}_x \mathbf{u}_\varepsilon$$

$$= \varepsilon \Delta B(\varrho_\varepsilon) - \varepsilon B''(\varrho_\varepsilon)|\nabla_x \varrho_\varepsilon|^2$$

for any function $B \in C^2[0,\infty)$ with B', B'' uniformly bounded. Moreover, since ϱ_ε satisfies the homogeneous Neumann boundary conditions and $\mathbf{u}_\varepsilon$ vanishes on $\partial\Omega$ in the sense of traces, we have

$$\partial_t B(\varrho_\varepsilon) + \mathrm{div}_x(B(\varrho_\varepsilon)\mathbf{u}_\varepsilon) + \Big(B'(\varrho_\varepsilon)\varrho_\varepsilon - B(\varrho_\varepsilon) \Big) \mathrm{div}_x \mathbf{u}_\varepsilon$$

$$= \varepsilon \mathrm{div}_x(1_\Omega \nabla_x B(\varrho_\varepsilon)) - \varepsilon 1_\Omega B''(\varrho)|\nabla_x \varrho_\varepsilon|^2 \text{ in } \mathcal{D}'((0,T) \times R^N) \qquad (7.90)$$

provided $B(0) = 0$.

If, in addition, B is convex, we deduce

$$\int_0^T \int_\Omega \psi \Big(B'(\varrho_\varepsilon)\varrho_\varepsilon - B(\varrho_\varepsilon) \Big) \mathrm{div}_x \mathbf{u}_\varepsilon \, \mathrm{d}x \, \mathrm{d}t$$

$$\leq \int_\Omega B(\varrho_{0,\delta}) \, \mathrm{d}x + \int_0^T \int_\Omega \psi_t B(\varrho_\varepsilon) \, \mathrm{d}x \, \mathrm{d}t$$

for any $\psi \in C^\infty[0,T]$, $\psi \geq 0$, $\psi(0) = 1$, $\psi(T) = 0$. Consequently, approximating $z \mapsto z \log(z)$ by a sequence of smooth convex functions we get

$$\int_0^T \int_\Omega \psi \varrho_\varepsilon \operatorname{div}_x \mathbf{u}_\varepsilon \, d\mathbf{x} \, dt \leq \int_\Omega \varrho_{0,\delta} \log(\varrho_{0,\delta}) \, d\mathbf{x} + \int_0^T \int_\Omega \psi_t \varrho_\varepsilon \log(\varrho_\varepsilon) \, d\mathbf{x} \, dt.$$

Here, the approximations of $z \log(z)$ can be taken, for instance, as

$$\mathcal{L}_k(\varrho) = z \int_1^\varrho \frac{\mathcal{T}_k(z)}{z^2} \, dz,$$

with non-decreasing functions $\mathcal{T}_k$

$$\mathcal{T}_k \in C^\infty(R^1), \quad \mathcal{T}_k(z) = 0 \ \text{ for } z \leq \frac{1}{k}, \quad \mathcal{T}_k(z) = k \quad \text{for } z \geq k,$$

$$\mathcal{T}_k(z) \to z \ \text{ uniformly on compacts in } (0, \infty).$$

Passing to the limit for $\varepsilon \to 0$ we obtain

$$\int_0^T \int_\Omega \psi \overline{\varrho \operatorname{div}_x \mathbf{u}} \, d\mathbf{x} \, dt \leq \int_\Omega \varrho_{0,\delta} \log(\varrho_{0,\delta}) \, d\mathbf{x} + \int_0^T \int_\Omega \psi_t \overline{\varrho \log(\varrho)} \, d\mathbf{x} \, dt,$$

from which we discover

$$\int_0^\tau \int_\Omega \overline{\varrho \operatorname{div}_x \mathbf{u}} \, d\mathbf{x} \leq \int_\Omega \varrho_{0,\delta} \log(\varrho_{0,\delta}) \, d\mathbf{x} - \int_\Omega \overline{\varrho \log(\varrho)}(\tau) \, d\mathbf{x} \qquad (7.91)$$

for any Lebesgue point τ of the function $\overline{\varrho \log(\varrho)}$—a weak limit of the sequence $\{\varrho_\varepsilon \log(\varrho_\varepsilon)\}_{\varepsilon > 0}$ in $L^\infty(0, T; L^q(\Omega))$, $q < \beta$.

On the other hand, one can use

$$\varphi(t, \mathbf{x}) = \psi(t)\eta(\mathbf{x}), \quad \psi \in \mathcal{D}(0, T), \ \psi \geq 0, \ \eta \geq 0, \ \eta \in \mathcal{D}(R^N), \ \eta|_\Omega = 1$$

as a test function for (7.89) to obtain

$$\int_0^\tau \int_\Omega \varrho \operatorname{div}_x \mathbf{u} \, d\mathbf{x} = \int_\Omega \varrho_{0,\delta} \log(\varrho_{0,\delta}) \, d\mathbf{x} - \int_\Omega \varrho \log(\varrho)(\tau) \, d\mathbf{x} \qquad (7.92)$$

for any $\tau \in [0, T]$. Note that, in accordance with Proposition 4.3, the function

$$t \mapsto \int_\Omega \varrho \log(\varrho)(t) \, d\mathbf{x}$$

is continuous on $[0, T]$.

Taking the sum of (7.91), (7.92) we arrive at inequality

$$\int_\Omega (\overline{\varrho \log(\varrho)} - \varrho \log(\varrho))(\tau)\,\mathrm{d}\mathbf{x}$$

$$\le \int_0^\tau \int_\Omega \overline{\varrho \operatorname{div}_x \mathbf{u}} - \varrho \operatorname{div}_x \mathbf{u}\,\mathrm{d}\mathbf{x}\,\mathrm{d}t \quad \text{for a.a. } \tau \in [0, T] \qquad (7.93)$$

where, by virtue of Lemma 7.7,

$$\int_Q \overline{\varrho \operatorname{div}_x \mathbf{u}} - \varrho \operatorname{div}_x \mathbf{u}\,\mathrm{d}\mathbf{x}\,\mathrm{d}t$$

$$\ge \frac{1}{\lambda + 2\mu} \liminf_{\varepsilon \to 0} \int_O \left(\overline{p_e(\varrho)\varrho} + \overline{\vartheta p_\vartheta(\varrho)\varrho} + \delta \overline{\varrho_\varepsilon^{\beta+1}} \right) - \overline{p}\varrho\,\mathrm{d}\mathbf{x}\,\mathrm{d}t$$

for any compact $O \subset ((0, T) \times \Omega)$. Here

$$\overline{p} = \overline{p_e(\varrho)} + \overline{\vartheta p_\vartheta(\varrho)} + \delta \overline{\varrho^\beta},$$

where, in accordance with our convention, the bar denotes a weak limit in $L^1(O)$.

Now, as the function $z \mapsto \delta z^\beta$ is increasing, we get

$$\liminf_{\varepsilon \to 0} \int_Q \varrho_\varepsilon^{\beta+1} - \overline{\varrho}^\beta \varrho\,\mathrm{d}\mathbf{x}\,\mathrm{d}t \ge 0. \qquad (7.94)$$

Furthermore, since nonlinear compositions of ϱ_ε satisfy (7.90), Lemma 6.3, together with estimate (7.70), yields

$$B(\varrho_\varepsilon) \to \overline{B(\varrho)} \text{ (strongly) in } L^2(0, T; W^{-1,2}(\Omega)) \qquad (7.95)$$

for any function $B \in C^2[0, \infty)$ with B', B'' uniformly bounded.

Thus we can suppose, in accordance with (7.81) that

$$\overline{\vartheta B(\varrho)} = \vartheta \overline{B(\varrho)}$$

for any smooth B as above, from which we easily deduce

$$\overline{\vartheta p_\vartheta(\varrho)} = \vartheta \overline{p_\vartheta(\varrho)} \text{ and } \overline{\vartheta p_\vartheta(\varrho)\varrho} = \vartheta \overline{p_\vartheta(\varrho)\varrho}.$$

In particular, as p_ϑ is a non-decreasing function of ϱ, we get

$$\overline{\vartheta p_\vartheta(\varrho)\varrho} = \vartheta \overline{p_\vartheta(\varrho)\varrho} \ge \vartheta \overline{p_\vartheta(\varrho)}\,\varrho = \overline{\vartheta p_\vartheta(\varrho)}\,\varrho.$$

Accordingly, relation (7.93) reduces to

$$\int_\Omega (\overline{\varrho \log(\varrho)} - \varrho \log(\varrho))(\tau)\,\mathrm{d}\mathbf{x} \le \frac{1}{\lambda + 2\mu} \int_0^\tau \int_\Omega \overline{p_e(\varrho)\varrho} - \overline{p_e(\varrho)}\varrho\,\mathrm{d}\mathbf{x}.$$

Now the elastic pressure p_e can be decomposed as in (6.61), and the same arguments as in Section 6.6, Part 6.6.3 can be used to conclude that

$$\overline{\varrho \log(\varrho)} = \varrho \log(\varrho) \text{ a.a. } \text{ on}(0, T) \times \Omega,$$

which is equivalent to strong convergence

$$\varrho_\varepsilon \to \varrho \text{ in } L^1((0,T) \times \Omega). \tag{7.96}$$

Consequently, the limit functions ϱ and $\mathbf{u}$ satisfy the momentum equation (1.35) in $\mathcal{D}'((0,T) \times \Omega)$.

7.4.4 Limit in the thermal energy equation

Since we have established for the sequence $\{\varrho_\varepsilon, \mathbf{u}_\varepsilon, \vartheta_\varepsilon\}_{\varepsilon>0}$ the same set of estimates as in Section 7.3, we can pass to the limit for $\varepsilon \to 0$ in the thermal energy inequality (7.72) exactly as in the proof of Proposition 7.5. Seeing that

$$\varepsilon \int_0^T \int_\Omega \nabla_x \varrho_\varepsilon \cdot \nabla_x \Big[\big(Q_\mathrm{h}(\vartheta_\varepsilon) - Q(\vartheta_\varepsilon) h(\vartheta_\varepsilon) \big) \varphi \Big] \, \mathrm{d}\mathbf{x} \, \mathrm{d}t \to 0$$

for any fixed test function φ (see (7.70), (7.81)), we obtain

$$\int_0^T \int_\Omega \Big((\delta + \varrho) Q_\mathrm{h}(\vartheta) \Big) \partial_t \varphi + \varrho Q_\mathrm{h}(\vartheta) \mathbf{u} \cdot \nabla_x \varphi + \mathcal{K}_\mathrm{h}(\vartheta) \Delta \varphi - \delta \vartheta^{\alpha+1} h(\vartheta) \varphi \, \mathrm{d}\mathbf{x} \, \mathrm{d}t$$

$$\leq \int_0^T \int_\Omega \Big((\delta - 1) h(\vartheta) \mathbb{S} : \nabla_x \mathbf{u} + \kappa(\vartheta) h'(\vartheta) |\nabla_x \vartheta|^2 \Big) \varphi \, \mathrm{d}\mathbf{x} \, \mathrm{d}t$$

$$+ \int_0^T \int_\Omega \vartheta h(\vartheta) p_\vartheta(\varrho) \operatorname{div}_x \mathbf{u} \varphi \, \mathrm{d}\mathbf{x} \, \mathrm{d}t - \int_\Omega (\varrho_{0,\delta} + \delta) Q_\mathrm{h}(\vartheta_{0,\delta}) \varphi(0) \, \mathrm{d}\mathbf{x} \tag{7.97}$$

for any h satisfying (4.48), and any test function φ as in (7.73).

The results obtained in this part are summarized in the following assertion.

Proposition 7.6 *Let $\Omega \subset R^N$ be a bounded domain of class $C^{2+\nu}$, $\nu > 0$. Moreover, let*

$$\beta > \max\{N, 4, \gamma\}, \quad and \quad \delta > 0$$

be given.

Then there exist approximate solutions $\varrho, \mathbf{u},$ and ϑ with the following properties:

(1) *The density ϱ is a non-negative function,*

$$\varrho \in C([0,T]; L_\mathrm{weak}^\beta(\Omega)), \quad \varrho(0) = \varrho_{0,\delta};$$

the velocity $\mathbf{u}$ belongs to the class $L^2(0,T; W_0^{1,2}(\Omega; R^N))$. The pair of functions $\varrho, \mathbf{u}$ is a renormalized solution of the continuity equation (1.34) in the sense Definition 4.1 on $(0,T) \times R^N$ provided they were extended to be zero outside Ω.

Moreover, the density ϱ belongs to space $L^{\beta+1}(O)$ for any compact $O \subset ((0,T) \times \Omega)$.

(2) *The functions ϱ, $\mathbf{u}$, and ϑ satisfy the momentum equation (1.35) in the sense of distributions, that is, in $\mathcal{D}'((0,T) \times \Omega)$, with*

$$p = p_e(\varrho) + \delta\varrho^\beta + \vartheta p_\vartheta(\varrho).$$

Moreover,

$$\varrho\mathbf{u} \in C([0,T]; L_{\text{weak}}^{2\gamma/\gamma+1}(\Omega; R^N))$$

satisfies the initial conditions

$$(\varrho\mathbf{u})(0) = \mathbf{m}_{0,\delta}.$$

(3) *The energy inequality*

$$\int_\Omega \varrho(\tau)\left(\frac{1}{2}|\mathbf{u}|^2 + P_e(\varrho) + \frac{\delta}{\beta-1}\varrho^{\beta-1}\right)(\tau) + (\delta + \varrho)Q(\vartheta)(\tau)\,\mathrm{dx}$$

$$+ \delta\int_0^\tau\int_\Omega \mathbb{S}:\nabla_x\mathbf{u} + \vartheta^{\alpha+1}\,\mathrm{dx}\,\mathrm{dt}$$

$$\leq \int_\Omega \frac{|\mathbf{m}_0|^2}{\varrho_{0,\delta}} + \varrho_{0,\delta}P_e(\varrho_{0,\delta}) + \frac{\delta}{\beta-1}\varrho_{0,\delta}^\beta + (\varrho_{0,\delta}+\delta)Q(\vartheta_{0,\delta})\,\mathrm{dx}$$

$$+ \int_0^\tau\int_\Omega \varrho\mathbf{f}\cdot\mathbf{u}\,\mathrm{dx}\,\mathrm{dt}$$

holds for a.a. $\tau \in [0,T]$.

(4) *The temperature ϑ is a non-negative function,*

$$\vartheta \in L^{\alpha+1}((0,T)\times\Omega), \nabla_x\vartheta^{\frac{\alpha+1-\omega}{2}} \text{ in } L^2((0,T)\times\Omega) \quad \textit{for any } \omega > 0,$$

satisfying the integral inequality (7.97).

Remark As ϱ is a renormalized solution of the continuity equation (1.34), the term

$$\int_0^\tau\int_\Omega p_b(\varrho)\,\mathrm{div}_x\mathbf{u}\,\mathrm{dx}\,\mathrm{dt}$$

appearing on the right-hand side of (7.72) is equal to

$$\int_\Omega \varrho_{0,\delta}P_b(\varrho_{0,\delta})\,\mathrm{dx} - \int_\Omega \varrho P_b(\varrho)(\tau)\,\mathrm{dx}$$

for any $\tau \in [0,T]$, where

$$P_b(\varrho) \equiv \int_1^\varrho \frac{p_b(z)}{z^2}\,\mathrm{dz}.$$

In accordance with hypothesis (3.1), the "bounded" component p_b of the elastic pressure can be taken such that $p_b \in C^2[0,\infty)$, $p_b(0) = 0$ and, consequently, the product $z \mapsto zP_b(z)$ is a continuous function on $[0,\infty)$ vanishing at zero.

7.5 Vanishing artificial pressure

Our ultimate goal in the proof of Theorem 7.1 is to carry out the limit process when the parameter δ tends to zero. Denote ϱ_δ, $\mathbf{u}_\delta$, ϑ_δ the corresponding *approximate solutions* constructed in Proposition 7.6.

7.5.1 Energy estimates

To begin with, we utilize the machinery assembled in Chapter 3 and in Chapter 5 to obtain uniform estimates independent of δ.

In light of the energy inequality stated in Proposition 7.6, it is easy to see

$$\varrho_\delta \text{ bounded in } L^\infty(0, T; L^\gamma(\Omega)), \tag{7.98}$$

$$\sqrt{\varrho_\delta}\mathbf{u}_\delta \text{ bounded in } L^\infty(0, T; L^2(\Omega, R^N)), \tag{7.99}$$

$$(\delta + \varrho_\delta)Q(\vartheta_\delta) \text{ bounded in } L^\infty(0, T; L^1(\Omega)), \tag{7.100}$$

and

$$\delta \int_0^T \int_\Omega \vartheta_\delta^{\alpha+1} \, \mathrm{dx}\, \mathrm{dt} \le c \tag{7.101}$$

where, by virtue of (7.17), the bounds are independent of $\delta > 0$. Note that

$$\left| \int_0^T \int_\Omega \varrho_\delta \mathbf{f} \cdot \mathbf{u}_\delta \, \mathrm{dx}\, \mathrm{dt} \right| \le \|\mathbf{f}\|_{L^\infty((0,T)\times\Omega)} \left(\frac{T}{2}M_\delta + \int_0^T \int_\Omega \frac{1}{2}\varrho_\delta |\mathbf{u}_\delta|^2 \, \mathrm{dx}\, \mathrm{dt} \right),$$

with the total mass $M_\delta \to M$ for $\delta \to 0$.

Now take

$$\varphi(t, \mathbf{x}) = \psi(t), \quad 0 \le \psi \le 1, \quad \psi \in \mathcal{D}(0, T), \quad h(\vartheta) = \frac{\omega}{\omega + \vartheta}, \quad \omega > 0,$$

in (7.97) to deduce

$$\int_0^T \int_\Omega \left(\frac{1-\delta}{\omega + \vartheta_\delta} \mathbb{S}_\delta : \nabla_x \mathbf{u}_\delta + \frac{\kappa(\vartheta_\delta)}{(\omega + \vartheta_\delta)^2}|\nabla_x \vartheta_\delta|^2 \right) \mathrm{dx}\, \mathrm{dt}$$

$$\le \int_0^T \int_\Omega \frac{\vartheta_\delta}{\omega + \vartheta_\delta} p_\vartheta(\varrho_\delta) \, \mathrm{div}_x \mathbf{u}_\delta \, \mathrm{dx}\, \mathrm{dt} + \delta \int_0^T \int_\Omega \vartheta_\delta^\alpha \, \mathrm{dx}\, \mathrm{dt}$$

$$- \int_\Omega (\varrho_{0,\delta} + \delta)Q_{\mathrm{h},\omega}(\vartheta_{0,\delta}) \, \mathrm{dx} + \int_\Omega (\varrho_\delta + \delta)Q_{\mathrm{h},\omega}(\vartheta_\delta)(T-) \, \mathrm{dx},$$

where

$$Q_{\mathrm{h},\omega}(\vartheta) = \int_1^\vartheta \frac{c_v(z)}{\omega + z} \, \mathrm{dz}.$$

Letting $\omega \to 0$ and taking hypothesis (1.38) together with the estimates (7.100), (7.101) into account, we deduce

$$\int_0^T \int_\Omega \frac{1}{1+\vartheta} (\mathbb{S}_\delta : \nabla_x \mathbf{u}_\delta) + |\nabla_x \vartheta_\delta|^2 + |\nabla_x \vartheta_\delta^{\alpha/2}|^2 \, \mathrm{d}\mathbf{x} \, \mathrm{d}t$$

$$\leq c \Big(1 + \int_0^T \int_\Omega p_\vartheta(\varrho_\delta) \, \mathrm{div}_x \mathbf{u}_\delta \, \mathrm{d}\mathbf{x} \, \mathrm{d}t \Big), \qquad (7.102)$$

with c independent of $\delta > 0$.

Now, utilizing estimate (7.98) together with hypothesis (3.2), we have

$$p_\vartheta(\varrho_\delta) \text{ bounded in } L^\infty(0,T;L^q(\Omega)) \text{ for a certain } q > 2 \text{ if } N = 2, \qquad (7.103)$$

or

$$p_\vartheta(\varrho_\delta) \text{ bounded in } L^\infty(0,T;L^N(\Omega)) \text{ for } N \geq 3. \qquad (7.104)$$

Thus in both cases one can use the fact that ϱ_δ is a renormalized solution of the continuity equation in order to get a bound on the integral appearing on the right-hand side of (7.102) which is independent of δ.

On the other hand, by virtue of Lemma 3.2,

$$\|\vartheta_\delta\|_{L^{2^*}(\Omega)} \leq c_1 \|\vartheta_\delta\|_{W^{1,2}(\Omega)} \leq c_2 \Big(\int_\Omega \varrho_\delta \vartheta_\delta \, \mathrm{d}\mathbf{x} + \|\nabla_x \vartheta_\delta\|_{L^2(\Omega)} \Big)$$

$$\leq c_3 \Big(\int_\Omega \varrho_\delta Q(\vartheta_\delta) \, \mathrm{d}\mathbf{x} + \|\nabla_x \vartheta_\delta\|_{L^2(\Omega)} \Big),$$

with the critical exponent 2^* given by (3.30). Consequently, relation (7.102) yields

$$\vartheta_\delta \text{ bounded in } L^2(0,T;W^{1,2}(\Omega)), \qquad (7.105)$$

together with

$$\nabla_x \vartheta_\delta^{\alpha/2} \text{ bounded in } L^2((0,T) \times \Omega). \qquad (7.106)$$

Furthermore, bootstraping (7.105), (7.106) we obtain

$$\vartheta_\delta^{\alpha/2} \text{ bounded in } L^2(0,T;W^{1,2}(\Omega)). \qquad (7.107)$$

Next, in accordance with Theorem 2.6, the bound (7.105) together with (7.103), (7.104) yield

$$\vartheta_\delta p_\vartheta(\varrho_\delta) \text{ bounded in } L^2((0,T) \times \Omega). \qquad (7.108)$$

With (7.108) at hand, we can repeat the same procedure taking now $h(\vartheta) = \frac{1}{(1+\vartheta)^\omega}$, $\omega \in (0,1)$.

To begin with, we let $\omega \to 0$ to obtain

$$\int_0^T \int_\Omega \mathbb{S}_\delta : \nabla_x \mathbf{u}_\delta \, \mathrm{d}\mathbf{x} \, \mathrm{d}t \leq c.$$

Thus we discover, in accordance with hypothesis (1.38), that

$$\mathbf{u}_\delta \text{ is bounded in } L^2(0, T; W_0^{1,2}(\Omega, R^N)). \tag{7.109}$$

Next, utilizing (7.109), we have

$$\int_0^T \int_\Omega \frac{1}{(1+\vartheta_\delta)^\omega} \mathbb{S}_\delta : \nabla_x \mathbf{u}_\delta + \omega \frac{\kappa(\vartheta_\delta)}{(1+\vartheta_\delta)^{1+\omega}} |\nabla_x \vartheta_\delta|^2 \, \mathrm{d}\mathbf{x} \, \mathrm{d}t \leq c,$$

where c is independent of both δ and ω, which yields

$$\|\vartheta_\delta^{(\alpha+1-\omega)/2}\|_{L^2(0,T;W^{1,2}(\Omega))} \leq c(\omega) \quad \text{for any } \omega > 0. \tag{7.110}$$

Finally, using the interpolation arguments as in Section 5.2 we establish an analogue of (5.26):

$$\int_{\{\varrho_\delta > \omega\}} \vartheta_\delta^{r(\alpha+1)} \, \mathrm{d}\mathbf{x} \, \mathrm{d}t \leq c(\omega) \quad \text{for a certain } r > 1. \tag{7.111}$$

7.5.2 *Temperature estimates*

Our aim now is to derive similar estimates on ϑ_δ to those in Section 5.2 in the region where $\varrho_\delta < \omega$, with ω a sufficiently small positive number.

Since the density ϱ_δ solves (1.34) in $\mathcal{D}'((0, T) \times R^N)$, the total mass M_δ is a contant of motion, and we have

$$\int_{\{\varrho_\delta(t) \geq \omega\}} \varrho_\delta(t) \, \mathrm{d}\mathbf{x} \geq M_\delta - \omega|\Omega| \geq \frac{M}{2} - \omega|\Omega|$$

in accordance with (7.14).

On the other hand, a straightforward application of Hölder's inequality (2.1) gives rise to

$$\int_{\{\varrho_\delta(t) \geq \omega\}} \varrho_\delta(t) \, \mathrm{d}\mathbf{x} \leq |\{\varrho_\delta \geq \omega\}|^{(\gamma-1)/\gamma} \|\varrho_\delta(t)\|_{L^\gamma(\Omega)}.$$

Consequently, by virtue of (7.98), there exists a function $d = d(\omega)$, which is independent of δ, such that

$$|\{\varrho_\delta(t) \geq \omega\}| \geq d(\omega) > 0 \quad \text{for all } t \in [0, T] \text{ provided } 0 \leq \omega < \frac{M}{2|\Omega|}. \quad (7.112)$$

Similarly to Section 5.2, we fix $0 < \omega < M/4|\Omega|$ and find a function $B \in C^\infty(R)$ such that

$$B : R \to R \text{ non-increasing}, \ B(z) = 0 \text{ for } z \leq \omega, \ B(z) = -1 \text{ for } z \geq 2\omega.$$

For each $t \in [0, T]$, let $\eta = \eta_\delta$ be the (unique) strong solution of the Neumann problem

$$\left\{ \begin{array}{c} \Delta \eta_\delta(t) = B(\varrho_\delta(t)) - \frac{1}{|\Omega|} \int_\Omega B(\varrho_\delta(t)) \, \mathrm{d}\mathbf{x} \text{ in } \Omega, \\[2mm] \nabla_x \eta_\delta \cdot \mathbf{n} = 0 \quad \text{on } \partial\Omega, \\[2mm] \int_\Omega \eta_\delta(t) \, \mathrm{d}\mathbf{x} = 0. \end{array} \right\} \quad (7.113)$$

Since the right-hand side of (7.113) is uniformly bounded independently of δ, there is a constant $\underline{\eta}$ such that

$$\eta_\delta(t, \mathbf{x}) \geq \underline{\eta} \quad \text{for all } t \in [0, T], \ \mathbf{x} \in \Omega, \ \delta > 0.$$

In accordance with Lemma 4.7, we can take a test function

$$\varphi(t, \mathbf{x}) \equiv \psi(t)(\eta_\delta(t, \mathbf{x}) - \underline{\eta}), \ \psi \in \mathcal{D}(0, T), \quad 0 \leq \psi \leq 1$$

in (7.97) to deduce

$$\int_0^T \int_\Omega \psi \mathcal{K}_\mathrm{h}(\vartheta_\delta) \Big(B(\varrho_\delta) - \frac{1}{|\Omega|} \int_\Omega B(\varrho_\delta) \, \mathrm{d}\mathbf{x} \Big) \, \mathrm{d}\mathbf{x} \, \mathrm{d}t$$

$$\leq 2 \|\eta_\delta\|_{L^\infty((0,T)\times\Omega)} \Big(\int_0^T \int_\Omega \delta \vartheta_\delta^{\alpha+1} + \vartheta_\delta |p_\vartheta(\varrho_\delta)| \, |\mathrm{div}_x \mathbf{u}_\delta| \, \mathrm{d}\mathbf{x} \, \mathrm{d}t \Big)$$

$$+ \|\nabla_x \eta_\delta\|_{L^\infty((0,T)\times\Omega)} \int_0^T \int_\Omega \varrho_\delta \mathcal{Q}_\mathrm{h}(\vartheta_\delta) |\mathbf{u}_\delta| \, \mathrm{d}\mathbf{x} \, \mathrm{d}t$$

$$+ \int_0^T \int_\Omega (\varrho_\delta + \delta) \mathcal{Q}_\mathrm{h}(\vartheta_\delta)(\underline{\eta} - \eta_\delta) \partial_t \psi - (\delta + \varrho_\delta) \mathcal{Q}_\mathrm{h}(\vartheta_\delta) \, \psi \partial_t \eta_\delta \, \mathrm{d}\mathbf{x} \, \mathrm{d}t. \quad (7.114)$$

Recall that any admissible function h is a non-increasing, non-negative function on $[0, \infty)$ with $h(0) = 1$.

Now we can take a sequence of functions $h = h_n \nearrow 1$ so that (7.114) gives rise to

$$\int_0^T \int_\Omega \psi \mathcal{K}(\vartheta_\delta)\left(B(\varrho_\delta) - \frac{1}{|\Omega|}\int_\Omega B(\varrho_\delta)\,\mathrm{d}\mathbf{x}\right)\mathrm{d}\mathbf{x}\,\mathrm{d}t$$

$$\leq 2\|\eta_\delta\|_{L^\infty((0,T)\times\Omega)}\left(\int_0^T\int_\Omega \delta\vartheta_\delta^{\alpha+1} + \vartheta_\delta|p_\vartheta(\varrho_\delta)|\,|\mathrm{div}_x\mathbf{u}_\delta|\,\mathrm{d}\mathbf{x}\,\mathrm{d}t\right)$$

$$+ \|\nabla_x\eta_\delta\|_{L^\infty((0,T)\times\Omega)}\int_0^T\int_\Omega \varrho_\delta Q(\vartheta_\delta)|\mathbf{u}_\delta|\,\mathrm{d}\mathbf{x}\,\mathrm{d}t$$

$$+ \int_0^T\int_\Omega (\delta + \varrho_\delta)Q(\vartheta_\delta)(\underline{\eta} - \eta_\delta)\partial_t\psi - (\varrho_\delta + \delta)Q(\vartheta_\delta)\,\psi\partial_t\eta_\delta\,\mathrm{d}\mathbf{x}\,\mathrm{d}t. \quad (7.115)$$

Moreover,

$$\int_0^T\int_\Omega \psi\mathcal{K}(\vartheta_\delta)\left(B(\varrho_\delta) - \frac{1}{|\Omega|}\int_\Omega B(\varrho_\delta)\,\mathrm{d}\mathbf{x}\right)\mathrm{d}\mathbf{x}\,\mathrm{d}t$$

$$= \int_{\{\varrho_\delta < \omega\}} \psi\mathcal{K}(\vartheta_\delta)\left(B(\varrho_\delta) - \frac{1}{|\Omega|}\int_\Omega B(\varrho_\delta)\,\mathrm{d}\mathbf{x}\right)\mathrm{d}\mathbf{x}\,\mathrm{d}t$$

$$+ \int_{\{\varrho_\delta \geq \omega\}} \psi\mathcal{K}(\vartheta_\delta)\left(B(\varrho_\delta) - \frac{1}{|\Omega|}\int_\Omega B(\varrho_\delta)\,\mathrm{d}\mathbf{x}\right)\mathrm{d}\mathbf{x}\,\mathrm{d}t$$

where, by virtue of (7.111), the second integral on the right-hand side is bounded independently of $\delta > 0$.

On the other hand,

$$-\frac{1}{|\Omega|}\int_\Omega B(\varrho_\delta)\,\mathrm{d}\mathbf{x} \geq -\frac{1}{|\Omega|}\int_{\{\varrho_\delta \geq 2\omega\}} B(\varrho_\delta)\,\mathrm{d}\mathbf{x} = \frac{|\{\varrho \geq 2\omega\}|}{|\Omega|} \geq \frac{d(2\omega)}{|\Omega|} > 0,$$

where we have used (7.112). Thus we get

$$\int_{\{\varrho < \omega\}} \psi\mathcal{K}(\vartheta_\delta)\left(B(\varrho_\delta) - \frac{1}{|\Omega|}\int_\Omega B(\varrho_\delta)\,\mathrm{d}\mathbf{x}\right)\mathrm{d}\mathbf{x}\,\mathrm{d}t$$

$$\geq \frac{d(2\omega)}{|\Omega|}\int_{\{\varrho_\delta < \omega\}} \psi\mathcal{K}(\vartheta_\delta)\,\mathrm{d}\mathbf{x}\,\mathrm{d}t.$$

This inequality, together with (7.111), (7.115), yields the desired estimates on ϑ_δ in the space $L^{\alpha+1}((0,T)\times\Omega)$ provided we show that the integrals on the right-hand side of (7.115) are bounded.

In view of estimates (7.98)–(7.109), it remains to show boundedness of the last integral

$$I = \int_0^T \int_\Omega (\delta + \varrho_\delta) Q(\vartheta_\delta)\, \psi\, \partial_t \eta_\delta\, \mathrm{d}\mathbf{x}\, \mathrm{d}t.$$

To this end, we use the fact that ϱ_δ is a solution of the renormalized continuity equation and, consequently,

$$\Delta \partial_t \eta_\delta = \partial_t (\Delta \eta_\delta) = \partial_t B(\varrho_\delta) - \frac{1}{|\Omega|} \int_\Omega \partial_t B(\varrho_\delta)\, \mathrm{d}\mathbf{x}$$

$$= -\mathrm{div}_x(B(\varrho_\delta)\mathbf{u}_\delta) - b(\varrho_\delta)\mathrm{div}_x\mathbf{u}_\delta + \frac{1}{|\Omega|} \int_\Omega b(\varrho_\delta)\mathrm{div}_x\mathbf{u}_\delta\, \mathrm{d}\mathbf{x},$$

where

$$b(\varrho) = B'(\varrho)\varrho - B(\varrho) \text{ is uniformly bounded.}$$

Similarly to Section 5.2, we deduce

$$\partial_t \eta \text{ in } L^2(0,T; W^{1,2}(\Omega))$$

which, together with (7.98), (7.105), yields boundedness of I.

Thus we have shown that

$$\vartheta_\delta \text{ is bounded in } L^{\alpha+1}((0,T) \times \Omega) \tag{7.116}$$

by a constant which is independent of $\delta > 0$.

7.5.3 Strict positivity of the temperature

It is easy to see that inequality (7.97) holds also for functions

$$h(\vartheta) = \frac{1}{\omega + \vartheta}, \quad \omega > 0,$$

$$\varphi(t, \mathbf{x}) = \psi(t), \quad 0 \leq \psi \leq 1, \quad \psi(0) = 1, \quad \psi(T) = 1, \quad \psi \in C^\infty[0,T].$$

Accordingly, in view of the estimates obtained above, we have

$$\int_0^T \int_\Omega (\delta + \varrho_\delta) Q_{\mathrm{h},\omega}(\vartheta_\delta)\partial_t\psi + \frac{k_1}{(\omega + \vartheta_\delta)^2}|\nabla\vartheta_\delta|^2\, \psi\, \mathrm{d}\mathbf{x}\, \mathrm{d}t$$

$$\leq c - \int_\Omega (\varrho_{0,\delta} + \delta) Q_{\mathrm{h},\omega}(\vartheta_{0,\delta})\, \mathrm{d}\mathbf{x},$$

where

$$Q_{\mathrm{h},\omega}(\vartheta) \equiv \int_1^\vartheta \frac{c_v(z)}{(\omega + z)}\, \mathrm{d}z.$$

Analogously to (3.26), letting $\omega \to 0$ we can use Lemma 3.2 to conclude that

$$\log(\vartheta_\delta) \text{ is bounded in } L^2((0,T) \times \Omega) \qquad (7.117)$$

by a constant independent of $\delta > 0$.

7.5.4 *Strong convergence of densities*

Having established all necessary estimates we address the question of convergence. As the approximate solutions $\varrho_\delta, \mathbf{u}_\delta,$ and ϑ_δ satisfy the equation of continuity (1.34), and the momentum equation (1.35) without any extra terms, the results obtained in Chapters 4, 6 can be used without modification.

It follows from estimates (7.98)–(7.101), (7.108) that the quantities ϱ_δ, $\mathbf{u}_\delta$, and

$$p_\delta \equiv p_e(\varrho_\delta) + \delta\varrho^\beta + \vartheta_\delta p_\vartheta(\varrho_\delta)$$

satisfy the hypotheses of Proposition 5.1. Consequently, we conclude that there is $\omega > 0$ such that

$$\int_O p_\delta \varrho_\delta^\omega \, \mathrm{d}\mathbf{x} \, \mathrm{d}t \le c(Q)$$

for any compact $O \subset ((0,T) \times \Omega)$. This implies local estimates

$$\|\varrho_\delta\|_{L^{\gamma+\omega}(O)} \le c(O), \qquad (7.118)$$

and

$$\delta \int_O \varrho_\delta^{\beta+\omega} \, \mathrm{d}\mathbf{x} \, \mathrm{d}t \le c(O) \qquad (7.119)$$

for any compact $O \subset ((0,T) \times \Omega)$.

In view of the above estimates, we may assume that, up to a subsequence,

$$\varrho_\delta \to \varrho \text{ in } C([0,T]; L^\gamma_{\text{weak}}(\Omega)),$$

$$\mathbf{u}_\delta \to \mathbf{u} \text{ weakly in } L^2(0,T; W_0^{1,2}(\Omega, R^N)),$$

and

$$\vartheta_\delta \to \vartheta \text{ weakly in } L^2(0,T; W^{1,2}(\Omega)), \qquad (7.120)$$

$$Q(\vartheta_\delta) \to \overline{Q(\vartheta)} \text{ weakly in } L^2(0,T; W^{1,2}(\Omega)).$$

Moreover, thanks to our choice of initial data $\varrho_{0,\delta}$,

$$\varrho(0,\mathbf{x}) = \varrho_0(\mathbf{x}) \text{ a.a. on } \Omega,$$

and

$$\delta \int_\Omega \frac{1}{\beta-1} \varrho_{0,\delta}^\beta \, \mathrm{d}\mathbf{x} \to 0 \quad \text{for } \delta \to 0.$$

As we have already observed, the space $L^\gamma(\Omega)$ is compactly imbedded into $W^{-1,2}(\Omega)$ and, consequently,

$$\varrho_\delta \mathbf{u}_\delta \to \varrho \mathbf{u} \text{ weakly(-*) in } L^\infty(0,T; L^{m_\infty}(\Omega; R^N)), \tag{7.121}$$

with

$$m_\infty \equiv \frac{2\gamma}{\gamma+1}.$$

Similarly,

$$\varrho_\delta Q(\vartheta_\delta) \to \overline{\varrho Q(\vartheta)} \text{ weakly in } L^2(0,T; L^{m_2}(\Omega)),$$

with m_2 as in (3.46).

The bound on $\partial_t(\varrho_\delta \mathbf{u}_\delta)$ resulting from (1.35) can be used to strengthen (7.121) to

$$\varrho_\delta \mathbf{u}_\delta \to \varrho \mathbf{u} \text{ in } C([0,T]; L^{m_\infty}_{\text{weak}}(\Omega; R^N)),$$

which yields, because of compact imbedding L^{m_∞} into $W^{-1,2}$, compactness of the convective terms:

$$\varrho_\delta \mathbf{u}_\delta \otimes \mathbf{u}_\delta \to \varrho \mathbf{u} \otimes \mathbf{u} \text{ weakly in } L^2(0,T; L^{c_2}(\Omega; R^{N^2})),$$

and

$$\varrho_\delta \mathbf{u}_\delta Q(\vartheta_\delta) \to \varrho \mathbf{u}\overline{Q(\vartheta)} \text{ weakly in } L^2(0,T; L^{c_2}(\Omega; R^{N^2}))$$

with $c_2 > 1$ as in Theorem 3.1. Moreover,

$$\varrho \mathbf{u}(0,\mathbf{x}) = \mathbf{m}_0(\mathbf{x}) \text{ a.a. on } \Omega.$$

In order to establish strong convergence of the sequence $\{\varrho_\delta\}_{\delta>0}$, we pursue the approach developed in Section 6.1. More specifically, a direct application of Proposition 6.1 yields

$$\lim_{\delta \to 0} \int_0^T \int_\Omega \psi\eta \Big(p_\delta - (\lambda + 2\mu) \operatorname{div}_x \mathbf{u}_\delta \Big) T_k(\varrho_\delta) \, \mathrm{d}\mathbf{x} \, \mathrm{d}t$$

$$= \lim_{\delta \to 0} \int_0^T \int_\Omega \psi\eta \Big(p_e(\varrho_\delta) + \vartheta_\delta p_\vartheta(\varrho_\delta) - (\lambda + 2\mu) \operatorname{div}_x \mathbf{u}_\delta \Big) T_k(\varrho_\delta) \, \mathrm{d}\mathbf{x} \, \mathrm{d}t$$

$$= \int_0^T \int_\Omega \psi\eta \Big(\overline{p_e} + \vartheta \overline{p_\vartheta(\varrho)} - (\lambda + 2\mu) \operatorname{div}_x \mathbf{u} \Big) \overline{T_k(\varrho)} \, \mathrm{d}\mathbf{x} \, \mathrm{d}t, \tag{7.122}$$

where, as usual,

$$p_e(\varrho_\delta) \to \overline{p_e(\varrho)} \text{ weakly in } L^{(\gamma+\omega)/\gamma}(O)$$

$$p_\vartheta(\varrho_\delta) \to \overline{p_\vartheta(\varrho)} \text{ in } C([0,T]; L^N_{\text{weak}}(\Omega)),$$

and

$$T_k(\varrho_\delta) \to \overline{T_k(\varrho)} \text{ in } C([0,T]; L^q_{\text{weak}}(\Omega)) \quad \text{for any } q \geq 1,$$

for any compact $O \subset (0,T) \times \Omega$. Note that

$$\delta \varrho_\delta^\beta \to 0 \text{ in } L^1((0,T) \times \Omega)$$

as a consequence of (7.119).

Similarly to Section 6.4, one can apply Proposition 6.2 in order to see that (7.122) implies boundedness of the *oscillations defect measure*

$$\mathbf{osc}_{\gamma+1}[\varrho_\delta \to \varrho](O) \leq c(|O|) \text{ for any compact } O \subset ((0,T) \times \Omega).$$

Consequently, in accordance with Proposition 6.3, the limit functions ϱ, $\mathbf{u}$ represent a *renormalized solution* of the continuity equation in the sense of Definition 4.1.

Finally, the results on *propagation of oscillations* stated in Section 6.6 can be now used directly in order to conclude that

$$\varrho_\delta \to \varrho \text{ strongly in } L^1((0,T) \times \Omega), \tag{7.123}$$

which can be improved, similarly to Section 6.7, to

$$\varrho_\delta \to \varrho \text{ in } C([0,T]; L^1(\Omega)). \tag{7.124}$$

Consequently, the limit functions ϱ, $\mathbf{u}$ satisfy the continuity equation

$$\partial_t \varrho + \text{div}_x(\varrho \mathbf{u}) = 0 \text{ in } \mathcal{D}'((0,T) \times R^N)$$

as well as in the sense of renormalized solutions. In other words, they represent a *variational solution* of equation (1.34) in the sense of Definition 4.2.

Similarly, the momentum equation

$$\partial_t(\varrho \mathbf{u}) + \text{div}_x(\varrho \mathbf{u} \otimes \mathbf{u}) + \nabla_x(p_e + \overline{\vartheta p_\vartheta}) = \text{div}_x \mathbb{S} + \varrho \mathbf{f}$$

is satisfied in $\mathcal{D}'((0,T) \times \Omega)$.

Finally, the limit quantities ϱ, $\mathbf{u}$, and ϑ satisfy the energy inequality

$$\int_\Omega \varrho(\tau) \left(\frac{1}{2} |\mathbf{u}|^2 + P_e(\varrho) + \overline{Q(\vartheta)} \right)(\tau) \, d\mathbf{x}$$

$$\leq \int_\Omega \frac{|\mathbf{m}_0|^2}{\varrho_0} + \varrho_0 P_e(\varrho_0) + \chi_0 \, d\mathbf{x} + \int_0^\tau \int_\Omega \varrho \mathbf{f} \cdot \mathbf{u} \, d\mathbf{x}$$

for a.a. $\tau \in [0,T]$, which can easily be verified using its "weak form" analogous to (7.74).

7.5.5   Thermal energy equation

In order to complete the proof of Theorem 7.1, we have to show that ϱ, $\mathbf{u}$, and ϑ represent a variational solution of the thermal energy equation (1.36) in the sense of Definition 4.5.

Approximating $h \nearrow 1$ in (7.97) we deduce

$$\int_0^T \int_\Omega \Big((\delta + \varrho_\delta) Q(\vartheta_\delta) \Big) \eta \partial_t \psi + \varrho_\delta Q(\vartheta_\delta) \mathbf{u}_\delta \cdot \nabla_x \eta \psi$$

$$+ \mathcal{K}(\vartheta_\delta) \psi \Delta \eta - \delta \vartheta_\delta^{\alpha+1} \eta \psi \, \mathrm{d}\mathbf{x} \, \mathrm{d}t \le \int_0^T \int_\Omega \vartheta_\delta p_\vartheta(\varrho_\delta) \, \mathrm{div}_x \mathbf{u}_\delta \, \eta \psi \, \mathrm{d}\mathbf{x} \, \mathrm{d}t$$

for any test functions

$$\psi, \eta \ge 0, \quad \psi \in \mathcal{D}(0, T), \quad \eta \in \mathcal{D}(\Omega).$$

Because of the uniform estimate (7.116), we are allowed to apply Lemma 6.3 to deduce

$$\varrho_\delta Q(\vartheta_\delta) \to \varrho \overline{Q(\vartheta)} \quad \text{in } L^2(0, T; W^{-1,2}(\Omega)).$$

In accordance with (7.120), we have

$$\varrho_\delta Q(\vartheta_\delta)^2 \to \varrho \overline{Q(\vartheta)}^2 \quad \text{in } \mathcal{D}'((0, T) \times \Omega).$$

Now relation (7.124) implies

$$\varrho Q(\vartheta_\delta)^2 \varphi = (\varrho - \varrho_\delta) Q(\vartheta_\delta)^2 \varphi + \varrho_\delta Q(\vartheta_\delta)^2 \varphi \to \varrho \overline{Q(\vartheta)}^2 \varphi,$$

and we infer, similarly to formula (6.79) of Section 6.7, that

$$\vartheta_\delta \to \overline{\vartheta} \text{ (strongly) in } L^r(\{\varrho > 0\}) \quad \text{for a certain } r > 1. \tag{7.125}$$

Now fixing h in (7.97) we can pass to limits for $\delta \to 0$ in the same way as in the proof of Theorem 6.2 to obtain

$$\int_0^T \int_\Omega \varrho Q_{\mathrm{h}}(\overline{\vartheta}) \partial_t \varphi + \varrho Q_{\mathrm{h}}(\overline{\vartheta}) \mathbf{u} \cdot \nabla_x \varphi + \overline{\mathcal{K}_{\mathrm{h}}(\vartheta)} \Delta \varphi \, \mathrm{d}\mathbf{x} \, \mathrm{d}t$$

$$\le \int_0^T \int_\Omega \Big(h(\overline{\vartheta}) \mathbb{S} : \nabla_x \mathbf{u} + h(\overline{\vartheta}) \, \overline{\vartheta} p_\vartheta(\varrho) \, \mathrm{div}_x \mathbf{u} \Big) \varphi \, \mathrm{d}\mathbf{x} \, \mathrm{d}t - \int_\Omega \varrho_0 Q_{\mathrm{h}}(\vartheta_0) \varphi(0) \, \mathrm{d}\mathbf{x},$$

$$\tag{7.126}$$

where

$$\varrho \overline{\mathcal{K}_{\mathrm{h}}(\vartheta)} = \varrho \mathcal{K}_{\mathrm{h}}(\overline{\vartheta}),$$

and

$$\|\overline{\mathcal{K}_h(\vartheta)}\|_{L^1((0,T)\times\Omega)} \leq c, \quad c \text{ independent of } h. \tag{7.127}$$

Note that

$$\delta\left|\int_0^T\int_\Omega \vartheta_\delta^{\alpha+1}h(\vartheta_\delta)\,\mathrm{d}\mathbf{x}\,\mathrm{d}t\right| \leq \delta\left|\int_{\{\vartheta_\delta<K\}}\vartheta^{\alpha+1}\,\mathrm{d}\mathbf{x}\,\mathrm{d}t\right| + h(K)\delta\int_0^T\int_\Omega \vartheta_\delta^{\alpha+1}\,\mathrm{d}\mathbf{x}\,\mathrm{d}t,$$

where the former integral on the right-hand side tends to zero for small δ while the latter can be made arbitrarily small taking K large enough in accordance with (7.116).

Next, we can use (7.117) together with convexity of the function $-\log$ to deduce

$$\log(\overline{\mathcal{K}_h(\vartheta)}) \in L^2((0,T)\times\Omega). \tag{7.128}$$

The idea now is the same as in the proof of Theorem 6.2, that is, we use the concept of a *renormalized limit* introduced in Section 6.8. Take

$$h(\vartheta) = \frac{1}{(1+\vartheta)^\omega}, \quad 0 < \omega < 1,$$

in (7.126), and let $\omega \to 0$ to obtain (4.41) for any test function φ as in (4.42). One has

$$\frac{1}{(1+\vartheta)^\omega} \nearrow 1 \quad \text{for } \omega \to 0,$$

and the monotone convergence theorem for the Lebesgue integral can be used. Note that, in accordance with (7.127),

$$\overline{\mathcal{K}_h(\vartheta)} \nearrow \overline{\mathcal{K}(\vartheta)},$$

with

$$\overline{\mathcal{K}(\vartheta)} \in L^1((0,T)\times\Omega),$$

where

$$\varrho\overline{\mathcal{K}(\vartheta)} = \varrho\mathcal{K}(\vartheta) \quad \text{on } (0,T)\times\Omega.$$

Finally, we set

$$\vartheta \equiv \mathcal{K}^{-1}(\overline{\mathcal{K}(\vartheta)}).$$

Obviously, the new function ϑ is non-negative, specifically, in accordance with (7.128),

$$\log(\vartheta) \in L^2((0,T)\times\Omega).$$

Moreover, ϑ satisfies (4.41) with

$$\varrho Q(\vartheta)(0) = \varrho_0 Q(\vartheta_0)$$

for any φ, and

$$\overline{\varrho\vartheta} = \varrho\vartheta \ \text{a.a.} \quad \text{on}\,(0,T)\times\Omega.$$

Theorem 7.1 has been proved.

7.6 Bibliographical notes

The first large data existence result for the *barotropic system* (1.34), (1.35) with pressure $p = p(\varrho)$ a non-decreasing function of the density was established by P.-L. Lions [85] on condition that

$$p(\varrho) \geq a\varrho^\gamma \text{ for large values of } \varrho,$$

where

$$\gamma \geq \frac{3}{2} \ \text{ if } N = 2, \quad \gamma \geq \frac{9}{5} \ \text{ if } N = 3, \quad \gamma > \frac{N}{2} \ \text{ for } N \geq 4.$$

We have already seen in Chapter 6 that these conditions are needed for the density ϱ to be square integrable. These restrictions were relaxed to $\gamma > \frac{N}{2}$, $N \geq 2$ in [45] and [50].

Surprisingly, the "simple" case of the *isothermal flow* where

$$p(\varrho) = \vartheta_0\varrho$$

remains open even for $N = 2$. The problem is the lack of estimates on the convective term $\varrho\mathbf{u}\otimes\mathbf{u}$. Some new ideas as well as positive existence results for the stationary flow were obtained by Plotnikov and Sokolowski [105].

The presentation in Chapter 7 follows [54]. Note that the existence problem for a *general* full system including the energy equation is far from being solved. The simplest and most interesting case of the ideal gas flow with

$$p(\varrho,\vartheta) = a\varrho\vartheta, \quad \lambda,\mu,\kappa \text{ constant}$$

is completely open. P.-L. Lions [85] gives a formal proof of weak stability under the additional hypothesis of boundedness of ϱ, $\mathbf{u}$, and ϑ in $L^\infty((0,T)\times\Omega)$.

There are many "small data" existence results represented by the pioneering paper by Matsumura and Nishida [92] we shall now briefly discuss. Consider the full system of the Navier–Stokes equations where

$$\mu = \mu(\varrho,\vartheta), \quad \lambda = \lambda(\varrho,\vartheta)$$

are smooth functions satisfying (1.38), the pressure is given by a general constitutive equation

$$p = p(\varrho,\vartheta), \quad \text{with } p, \frac{\partial p}{\partial \varrho}, \frac{\partial p}{\partial \vartheta} > 0,$$

and

$$\kappa = \kappa(\varrho,\vartheta), \quad \kappa > 0.$$

Consider the no-slip boundary conditions (1.41) for the velocity **u** together with the Dirichlet boundary conditions

$$\vartheta|_{\partial\Omega} = b,$$

where $\Omega \subset R^3$ is a regular domain with smooth boundary. Matsumura and Nishida showed the following result (Theorem 2.1 in [92]).

Theorem 7.2 *Let $\Omega \subset R^3$ be a domain with compact and smooth boundary. Let the quantities μ, λ, p, and κ comply with the hypotheses stated above. Moreover, let the initial data ϱ_0, $\mathbf{u}_0$, ϑ_0 belong to the Sobolev space $W^{3,2}(\Omega)$ and satisfy the compatibility conditions*

$$
\left\{
\begin{array}{c}
\mathbf{u}_0 = 0, \quad \theta_0 = b, \\[2mm]
\frac{\partial}{\partial x_i} p(\varrho_0, \vartheta_0) = \frac{\partial}{\partial x_j}\left(\mu(\varrho_0, \vartheta_0)\frac{\partial u_0^i}{\partial x_j}\right) + \frac{\partial}{\partial x_i}\left((\lambda(\varrho_0, \vartheta_0)\right. \\[2mm]
\left. +\mu(\varrho_0, \vartheta_0))(\frac{\partial u_0^j}{\partial x_j})\right) - \varrho_0 f^i, \\[2mm]
p(\varrho_0, \vartheta_0)\frac{\partial u_0^j}{\partial x_j} = \frac{\partial}{\partial x_j}\left(\kappa(\varrho_0, \vartheta_0)\frac{\partial \theta_0}{\partial x_j}\right) + \frac{\mu(\varrho_0, \vartheta_0)}{2}\left(\frac{\partial u_0^j}{\partial x_k} + \frac{\partial u_0^k}{\partial x_j}\right)^2 \\[2mm]
+\lambda(\varrho_0, \vartheta_0)\left(\frac{\partial u_0^j}{\partial x_j}\right)^2 \\[2mm]
i = 1, 2, 3
\end{array}
\right\}
$$

on the boundary $\partial\Omega$. Finally, let $f^i = \partial F/\partial x_i$, $i = 1, 2, 3$ where F belongs to the Sobolev space $W^{5,2}(\Omega)$.

Then there exists $\varepsilon > 0$ such that the initial-boundary value problem for the Navier–Stokes system (1.34)–(1.36) posseses a unique solution ϱ, $\mathbf{u}$, ϑ on the time interval $(0, \infty)$ provided the initial data satisfy

$$\|\varrho_0 - \overline{\varrho}\|_{W^{3,2}(\Omega)} + \|\mathbf{u}_0\|_{W^{3,2}(\Omega)} + \|\vartheta_0 - b\|_{W^{3,2}(\Omega)} + \|F\|_{W^{5,2}(\Omega)} < \varepsilon$$

where

$$\overline{\varrho} = \frac{1}{|\Omega|}\int_\Omega \varrho_0\, dx.$$

Related results may be found in [63], [64], [65], [120], [25] among others.

Several concepts of generalized solutions have been developed to accommodate the lack of compactness as well as of *a priori* estimates for the full system. Many of these techniques are reminiscent of the theory of *measure-valued* solutions initiated by DiPerna [29] and further developed in [89]. One possible approach is briefly discussed in chapter 8 of [85], where the case

$$p(\vartheta, \varrho) \approx (\vartheta + \delta)\varrho^a, \quad \kappa(\vartheta) \approx \vartheta^b, \quad \delta \geq 0,$$

is considered with a and b "large enough". The weak formulation takes into account both the energy and the entropy equation discussed in Chapter 1. These contain several terms replacing certain composed functions of the state variables, which either vanish or are equal to the "right" expressions provided we know the weak solution is smooth. Such a formulation still satisfies the basic principle of compatibility, that is, "weak+smooth" implies "classical" solution but, on the other hand, the solutions are weaker than the "classical" distributional solutions.

To conclude, let us mention several attempts to show that global classical solutions either for the barotropic or the full system in fact *cannot* exist. Xin [124] considered the full system (1.34)–(1.36) on R^3 with *compactly supported* initial distribution of the density. He showed that there is no global in-time solution belonging to the space $C^1([0, \infty); W^{m,2}(R^3))$ for $m > 3$. However, such a result does not seem to solve (in a negative way) the question of regularity for compressible flows as the Navier–Stokes system represents a mathematical model of non-dilute fluids for which the density should be bounded below away from zero. It is quite natural, therefore, to expect the problem to be ill posed when vacuum regions are allowed to appear at the initial time.

Another attempt to construct a counter-example to global existence was done by Vaigant [118] and Desjardins [27]. The singular solution they construct solves the problem with a suitable *unbounded* external force $\mathbf{f}$. Here again, the weakness of this approach stems from the necessity of using a singular forcing term to make solutions "explode" in a finite time.

BIBLIOGRAPHY

[1] R. A. Adams. *Sobolev spaces.* Academic Press, New York, 1975.

[2] S. Agmon, A. Douglis, and L. Nirenberg. Estimates near the boundary for solutions of elliptic partial differential equations. *Commun. Pure Appl. Math.*, **12**: 623–727, 1959.

[3] R. Alexandre and C. Villani. On the Boltzmann equation for long-range interactions. *Commun. Pure Appl. Math.*, **55**: 30–70, 2002.

[4] H. Amann. *Linear and quasilinear parabolic problems, I.* Birkhäuser Verlag, Basel, 1995.

[5] S. N. Antontsev, A. V. Kazhikhov, and V. N. Monakhov. *Krajevyje zadaci mechaniki neodnorodnych zidkostej.* Novosibirsk, 1983.

[6] J. M. Ball. A version of the fundamental theorem for Young measures. *In Lecture Notes in Physics 344*, Springer-Verlag, pp. 207–215, 1989.

[7] J. M. Ball and F. Murat. Remarks on Chacons biting lemma. *Proc. Amer. Math. Soc.*, **107**: 655–663, 1989.

[8] C. Bardos and B. Nicolaenko. Navier–Stokes equations and dynamical systems. In *Handbook of Dynamical Systems*, Vol. 2. B. Fiedler (ed.). North-Holland, Amsterdam, pp. 503–598, 2002.

[9] G. K. Batchelor. *An introduction to fluid dynamics.* Cambridge University Press, Cambridge, 1967.

[10] E. Becker. *Gasdynamik.* Teubner-Verlag, Stuttgart, 1966.

[11] L. Boccardo and F. Murat. Almost everywhere convergence of the gradients of solutions to elliptic and parabolic equations. *Nonlinear Anal.*, **19**(6): 581–597, 1992.

[12] M. E. Bogovskii. Solution of some vector analysis problems connected with operators div and grad (in Russian). *Trudy Sem. S. L. Sobolev*, **80**(1): 5–40, 1980.

[13] W. Borchers and H. Sohr. On the equation $rotv = g$ and $divu = f$ with zero boundary conditions. *Hokkaido Math. J.*, **19**: 67–87, 1990.

[14] H. Brezis. *Opérateurs maximaux monotones et semi-groupes de contractions dans les espaces de Hilbert.* North-Holland, Amsterdam, 1973.

[15] P. W. Bridgeman. *The physics of high pressure.* Dover Publications, New York, 1970.

[16] J. K. Brooks and R. V. Chacon. Continuity and compactness of measures. *Adv. Math.*, **37**: 16–26, 1980.

[17] L. Caffarelli, R. V. Kohn, and L. Nirenberg. On the regularity of the solutions of the Navier–Stokes equations. *Commun. Pure Appl. Math.*, **35**: 771–831, 1982.

[18] A. P. Calderón and A. Zygmund. On singular integrals. *Amer. J. Math.*, **78**: 289–309, 1956.

[19] T. Cazenave and A. Haraux. *An introduction to semilinear evolution equations.* Oxford University Press, Oxford, 1998.

[20] S. Chapman and T. G. Cowling. *Mathematical theory of non-uniform gases.* Cambridge University Press, Cambridge, 1990.

[21] H. J. Choe and B. J. Jin. Regularity of weak solutions of the compressible Navier–Stokes equations. 1999. Preprint.

[22] A. J. Chorin and J. E. Marsden. *A mathematical introduction to fluid mechanics.* Springer-Verlag, New York, 1979.

[23] R. Coifman and Y. Meyer. On commutators of singular integrals and bilinear singular integrals. *Trans. Amer. Math. Soc.*, **212**: 315–331, 1975.

[24] C. M. Dafermos. *Hyperbolic conservation laws in continuum physics.* Springer-Verlag, Berlin, 2000.

[25] R. Danchin. Global existence in critical spaces for compressible Navier–Stokes equations. *Inv. Math.*, **141**: 579–614, 2000.

[26] P. Degond and M. Lemou. On the viscosity and thermal conduction of fluids with multivalued internal energy. *Euro. J. Mech. B-Fluids*, **20**: 303–327, 2001.

[27] B. Desjardins. On weak solutions of the compressible isentropic Navier–Stokes equations. *Appl. Math. Lett.*, **12**: 107–111, 1999.

[28] R. J. DiPerna. Convergence of the viscosity method for isentropic gas dynamics. *Commun. Math. Phys.*, **91**: 1–30, 1983.

[29] R. J. DiPerna. Measure-valued solutions to conservation laws. *Arch. Rat. Mech. Anal.*, **88**: 223–270, 1985.

[30] R. J. DiPerna and P.-L. Lions. On the Fokker–Planck–Boltzmann equation. *Commun. Math. Phys.*, **120**: 1–23, 1988.

[31] R. J. DiPerna and P.-L. Lions. On the Cauchy problem for Boltzmann equations: global existence and weak stability. *Ann. Math.*, **130**: 312–366, 1989.

[32] R. J. DiPerna and P.-L. Lions. Ordinary differential equations, transport theory and Sobolev spaces. *Invent. Math.*, **98**: 511–547, 1989.

[33] R. J. DiPerna and A. Majda. Reduced Hausdorff dimension and concentration cancellation for two-dimensional incompressible flow. *J. Amer. Math. Soc.*, **1**: 59–95, 1988.

[34] B. Ducomet. Global existence for a simplified model of nuclear fluid in one space dimension. *J. Math. Fluid Mech.*, **2**: 1–15, 2000.

[35] B. Ducomet. Simplified models of quantum fluids of nuclear physics. *Math. Bohem.*, **126**: 323–336, 2001.

[36] B. Ducomet, E. Feireisl, H. Petzeltová, and I. Straškraba. Existence global pour un fluide barotrope autogravitant. *C. R. Acad. Sci. Paris, Sér. I.*, **332**: 627–632, 2001.

[37] N. Dunford and J. Schwartz. *Linear operators.* Interscience, New York, 1958.

[38] D. B. Ebin. Viscous fluids in a domain with frictionless boundary. In *Global Analysis—Analysis on Manifolds*, H. Kurke, J. Mecke, H. Triebel, R. Thiele (eds). Teubner-Texte zur Mathematik 57, Teubner, Leipzig, pp. 93–110, 1983.

[39] R. E. Edwards. *Functional analysis*. Holt-Rinehart-Winston, New York, 1965.

[40] I. Ekeland and R. Temam. *Convex analysis and variational problems*. North-Holland, Amsterdam, 1976.

[41] L. C. Evans. *Weak convergence methods for nonlinear partial differential equations*. Amer. Math. Soc., Providence, 1990.

[42] L. C. Evans and R. F. Gariepy. *Measure theory and fine properties of functions*. CRC Press, Boca Raton, 1992.

[43] H. Fan and M. Slemrod. Dynamic flows with liquid/vapor phase transitions. *Handbook of mathematical fluid dynamics*, Vol. I, Chapter 4. Elsevier Science, Amsterdam, pp. 373–420, 2002.

[44] E. Feireisl. Global attractors for the Navier–Stokes equations of three-dimensional compressible flow. *C.R. Acad. Sci. Paris, Sér. I*, **331**: 35–39, 2000.

[45] E. Feireisl. On compactness of solutions to the compressible isentropic Navier–Stokes equations when the density is not square integrable. *Comment. Math. Univ. Carolinae*, **42**(1): 83–98, 2001.

[46] E. Feireisl. Recent progress in the mathematical theory of viscous compressible fluids. In *Mathematical fluid dynamics—Recent results and open questions*. Birkhäuser, Basel, pp. 73–104, 2001.

[47] E. Feireisl. Compressible Navier–Stokes equations with a non-monotone pressure law. *J. Diff. Eqns*, **184**: 97–108, 2002.

[48] E. Feireisl. On the motion of rigid bodies in a viscous compressible fluid. *Arch. Rational Mech. Anal.*, **167**: 281–308, 2003.

[49] E. Feireisl. Shape optimization in viscous compressible fluids. *Appl. Math. Optim.*, **47**: 59–78, 2003.

[50] E. Feireisl, A. Novotný, and H. Petzeltová. On the existence of globally defined weak solutions to the Navier–Stokes equations of compressible isentropic fluids. *J. Math. Fluid Dynam.*, **3**: 358–392, 2001.

[51] E. Feireisl and H. Petzeltová. Large-time behaviour of solutions to the Navier–Stokes equations of compressible flow. *Arch. Rational Mech. Anal.*, **150**: 77–96, 1999.

[52] E. Feireisl and H. Petzeltová. On integrability up to the boundary of the weak solutions of the Navier–Stokes equations of compressible flow. *Commun. Partial Diff. Eqns*, **25**(3–4): 755–767, 2000.

[53] E. Feireisl and H. Petzeltová. Asymptotic compactness of global trajectories generated by the Navier–Stokes equations of compressible fluid. *J. Diff. Eqns*, **173**: 390–409, 2001.

[54] E. Feireisl and H. Petzeltová. Global existence for the full system of the Navier–Stokes equations of a viscous heat conducting fluid. *Arch. Rational Mech. Anal.*, 2003 (submitted).

[55] A. Friedman. *Partial differential equations of parabolic type.* Prentice-Hall, Englewood Cliffs, NJ, 1964.

[56] A. Friedman. *Variational principles and free-boundary problems.* John Wiley, New York, 1982.

[57] H. Gajewski, K. Gröger, and K. Zacharias. *Nichtlineare Operatorgleichungen und Operatordifferentialgleichungen.* Akademie Verlag, Berlin, 1974.

[58] G. P. Galdi. *An introduction to the mathematical theory of the Navier–Stokes equations, I.* Springer-Verlag, New York, 1994.

[59] G. Gallavotti. *Foundations of fluid dynamics.* Springer-Verlag, New York, 2002.

[60] P. Gerard. Microlocal defect measures. *Commun. Partial Diff. Eqns*, **16**: 1761–1794, 1991.

[61] R. A. Granger. *Fluid mechanics.* Dover Publications, New York, 1995.

[62] A. E. Green and R. S. Rivlin. Theories of elasticity with stress multipoles. *Arch. Rational Mech. Anal.*, **17**: 85–112, 1964.

[63] D. Hoff. Global solutions of the Navier–Stokes equations for multidimensional compressible flow with discontinuous initial data. *J. Diff. Eqns*, **120**: 215–254, 1995.

[64] D. Hoff. Strong convergence to global solutions for multidimensional flows of compressible, viscous fluids with polytropic equations of state and discontinuous initial data. *Arch. Rational Mech. Anal.*, **132**: 1–14, 1995.

[65] D. Hoff. Discontinuous solutions of the Navier–Stokes equations for multidimensional flows of heat conducting fluids. *Arch. Rational Mech. Anal.*, **139**: 303–354, 1997.

[66] D. Hoff and J. Smoller. Non-formation of vacuum states for Navier–Stokes equations. *Comm. Math. Phys.*, **216**: 255–276, 2001.

[67] S. Jiang and P. Zhang. On spherically symmetric solutions of the compressible isentropic Navier–Stokes equations. *Commun. Math. Phys.*, **215**: 559–581, 2001.

[68] L. V. Kantorowich and G. P. Akilov. *Functional Analysis (in Russian).* Nauka, Moscow, 1977.

[69] J. L. Kelley. *General topology.* Van Nostrand, Inc., Princeton, 1957.

[70] R. Kippenhahn and A. Weigert. *Stellar structure and evolution.* Springer-Verlag, New York, 1994.

[71] H. Kozono and H. Sohr. Regularity criterion of weak solutions to the Navier–Stokes equations. *Adv. Diff. Eqns*, **2**: 535–554, 1997.

[72] P. Kruus. *Liquids and solutions.* Marcel Dekker, New York, 1977.

[73] S. N. Kruzhkov. First order quasilinear equations in several space variables (in Russian). *Math. Sbornik*, **81**: 217–243, 1970.

[74] J. Kurzweil. *Ordinary differential equations.* Elsevier, Amsterdam, 1986.

[75] O. A. Ladyzhenskaya, V. A. Solonnikov, and N. N. Uraltseva. *Linear and quasilinear equations of parabolic type.* Trans. Math. Monograph 23, Amer. Math. Soc., Providence, 1968.

[76] H. Lamb. *Hydrodynamics.* Cambridge University Press, Cambridge, 1932.

[77] L. D. Landau and E. M. Lifshitz. *Fluid mechanics.* Pergamon, New York, 1968.

[78] J. Leray. Sur le mouvement d'un liquide visqueux emplissant l'espace. *Acta Math.*, **63**: 193–248, 1934.

[79] A. L. Letter and J. D. Walecka. *Quantum theory of many-particle systems.* McGraw Hill, New York, 1971.

[80] J. Lighthill. *Waves in fluids.* Cambridge University Press, Cambridge, 1978.

[81] J.-L. Lions. *Quelques méthodes de résolution des problèmes aux limites non linéaires.* Dunod, Gautthier-Villars, Paris, 1969.

[82] J.-L. Lions and E. Magenes. *Problèmes aux limites non homogènes et applications. I.* Dunod, Gautthier-Villars, Paris, 1968.

[83] P.-L. Lions. Compacité des solutions des équations de Navier–Stokes compressible isentropiques. *C.R. Acad. Sci. Paris, Sér I.*, **317**: 115–120, 1993.

[84] P.-L. Lions. *Mathematical topics in fluid dynamics, Vol. 1, Incompressible models.* Oxford Science Publication, Oxford, 1996.

[85] P.-L. Lions. *Mathematical topics in fluid dynamics, Vol. 2, Compressible models.* Oxford Science Publication, Oxford, 1998.

[86] P.-L. Lions. Bornes sur la densité pour les équations de Navier–Stokes compressible isentropiques avec conditions aux limites de Dirichlet. *C.R. Acad. Sci. Paris, Sér I.*, **328**: 659–662, 1999.

[87] P.-L. Lions, B. Perthame, and E. Souganidis. A kinetic formulation of multidimensional scalar conservation laws and related equations. *J. Amer. Math. Soc.*, **7**(1): 169–191, 1994.

[88] A. Lunardi. *Analytic semigroups and optimal regularity in parabolic problems.* Birkhäuser, Berlin, 1995.

[89] J. Málek, J. Nečas, M. Rokyta, and M. Růžička. *Weak and measure-valued solutions to evolutionary PDE's.* Chapman and Hall, London, 1996.

[90] A. E. Mamontov. Global solvability of the multidimensional Navier–Stokes equations of a compressible fluid with nonlinear viscosity, I. *Siberian Math. J.*, **40**(2): 351–362, 1999.

[91] A. E. Mamontov. Global solvability of the multidimensional Navier–Stokes equations of a compressible fluid with nonlinear viscosity, II. *Siberian Math. J.*, **40**(3): 541–555, 1999.

[92] A. Matsumura and T. Nishida. The initial value problem for the equations of motion of compressible and heat conductive fluids. *Commun. Math. Phys.*, **89**: 445–464, 1983.

[93] Š. Matušů-Nečasová and A. Novotný. Measure-valued solution for non-newtonian compressible isothermal monopolar fluid. *Acta Appl. Math.*, **37**: 109–128, 1994.

[94] V. G. Maz'ya. *Prostranstva S. L. Soboleva (in Russian)*. Izdatelstvo Leningradskogo Universiteta, Leningrad, 1985.

[95] R. E. Meyer. *Introduction to mathematical fluid dynamics*. Wiley, New York, 1971.

[96] B. Mihalas and B. Weibel-Mihalas. *Foundations of radiation hydrodynamics*. Dover Publications, Dover, 1984.

[97] J. Nečas. *Les méthodes directes en théorie des équations elliptiques*. Academia, Praha, 1967.

[98] J. Nečas, A. Novotný, and M. Šilhavý. Global solution to the ideal compressible heat-conductive multipolar fluid. *Comment. Math. Univ. Carolinae*, **30**: 551–564, 1989.

[99] J. Nečas, A. Novotný, and M. Šilhavý. Global solutions to the compressible isothermal multipolar fluid. *J. Math. Anal. Appl.*, **162**: 223–241, 1991.

[100] J. Nečas and M. Šilhavý. Viscous multipolar fluids. *Quart. Appl. Math.*, **49**: 247–266, 1991.

[101] L. Nirenberg. *Topics in nonlinear functional analysis*. Courant Institute, New York, 1974.

[102] A. Novotný and I. Straškraba. Convergence to equilibria for compressible Navier–Stokes equations with large data. *Ann. Mat. Pura Appl.*, **179**: 263–287, 2001.

[103] A. Novotný and I. Straškraba. Stabilization of weak solutions to compressible Navier–Stokes equations. *J. Math. Kyoto Univ.*, **40**: 217–245, 2000.

[104] B. Opic and A. Kufner. *Hardy-type inequalities*. Longman, Pitman Res. Notes in Math. 219, Essex, 1990.

[105] P. I. Plotnikov and J. Sokolowski. On compactness domain dependence and existence of steady state solutions to compressible isothermal Navier–Stokes equations. 2002. Preprint.

[106] M. H. Protter and H. F. Weinberger. *Maximum principles in differential equations*. Prentice Hall, Inc., London, 1967.

[107] J. Quastel and S. R. S. Varadhan. Diffusion semigroups and diffusion processes corresponding to degenerate divergence form operators. *Commun. Pure Appl. Math.*, **50**(7): 667–706, 1997.

[108] D. Serre. Variation de grande amplitude pour la densité d'un fluid viscueux compressible. *Phys. D*, **48**: 113–128, 1991.

[109] J. Simon. Compact sets and the space $L^p(0, T; B)$. *Ann. Mat. Pura Appl.*, **146**: 65–96, 1987.

[110] H. H. K. Tang and C. Y. Wong. Exactly central heavy-ions collisions by nuclear hydrodynamics. *Phys. Rev. C*, **21**: 1846–1863, 1980.

[111] L. Tartar. Compensated compactness and applications to partial differential equations. *Nonlinear Anal. and Mech., Heriot-Watt Sympos.*, L. J. Knopps (ed.), Research Notes in Math 39. Pitman, Boston, pp. 136–211, 1975.

[112] L. Tartar. H-measures, a new approach for studying homogenization, oscillations and concentration effects in partial differential equations. *Proc. Royal Soc. Edinburgh*, **115A**: 193–230, 1990.

[113] H. Triebel. *Interpolation theory, function spaces, differential operators.* VEB Deutscher Verlag der Wissenschaften, Berlin, 1978.

[114] H. Triebel. *Theory of function spaces.* Geest and Portig K.G., Leipzig, 1983.

[115] C. Truesdell. *The elements of continuum mechanics.* Springer-Verlag, New York, 1966.

[116] C. Truesdell. *A first course in rational continuum mechanics.* Academic Press, New York, 1991.

[117] C. Truesdell and K. R. Rajagopal. *An introduction to the mechanics of fluids.* Birkhäuser, Boston, 2000.

[118] V. A. Vaigant. An example of the nonexistence with respect to time of the global solutions of Navier–Stokes equations for a compressible viscous barotropic fluid (in Russian). *Dokl. Akad. Nauk*, **339**(2): 155–156, 1994.

[119] V. A. Vaigant and A. V. Kazhikhov. On the existence of global solutions to two-dimensional Navier–Stokes equations of a compressible viscous fluid (in Russian). *Sibirskij Mat. Z.*, **36**(6): 1283–1316, 1995.

[120] A. Valli and M. Zajaczkowski. Navier–Stokes equations for compressible fluids: Global existence and qualitative properties of the solutions in the general case. *Commun. Math. Phys.*, **103**: 259–296, 1986.

[121] G. J. Van Wylen and R. E. Sonntag. *Fundamentals of classical thermodynamics.* John Wiley, New York, 1985.

[122] C. Villani. A review of mathematical topics in collisional kinetic theory. In *Handbook of Mathematical Fluid Dynamics*, Vol. I, Chapter 2. Elsevier Sciences, North-Holland, Amsterdam, pp. 71–305, 2002.

[123] C. Y. Wong. Comparison of nuclear hydrodynamics and time dependent Hartree–Fock results. *Phys. Lett.*, **66B**: 19–22, 1977.

[124] Z. Xin. Blowup of smooth solutions to the compressible Navier–Stokes equation with compact density. *Commun. Pure Appl. Math.*, **51**: 229–240, 1998.

[125] K. Yosida. *Functional analysis.* Springer-Verlag, New York, 1980.

[126] Y. B. Zel'dovich and Y. P. Raizer. *Physics of shock waves and high-temperature hydrodynamic phenomena.* Academic Press, New York, 1966.

[127] W. P. Ziemer. *Weakly differentiable functions.* Springer-Verlag, New York, 1989.

INDEX